THE POISONOUS PLANTS OF BOMBAY

Other Related Books

Alpine Flora of Kashmir Himalaya; *U. Dhar & P. Kachroo*
Bibliography on Indian Ethnobotany 2002; *S.K. Jain*
Bioclimatology and Plant Geography of Peninsular India; *V.M. Meher-Homji*
Biodiversity Endangered; *A.B. Chaudhuri & D.D. Sarkar*
Biodiversity and Vegetation of Pachmarhi Hill; *V.P. Singh*
Biodiversity, Taxonomy & Ecology; *R.K. Tandon & Prithipalsingh*
Bonsai an Art - *Deshbeer Singh*
Central Asia and Kashmir Himalaya: Archaeobotany and Floristics; *P. Kachroo*
Contribution to Indian Ethnobotany (3rd Ed.); *S.K. Jain (Ed.)*
The Economic Plants of North-East India; *Sudhir Kumar*
Environment and Adaptive Biology of Plans; *D.D. Chawan (Ed.)*
Ethnobiology; *R.K. Sinha*
Ethnobotany and Medicinal Plants of Indian Subcontinent; *J.K. Maheshwari*
Ethnobotany of Rajasthan (India); *V. Singh & R.P. Pandey*
Ethnobotany of South Asia; *J.K. Maheshwari*
Flora of Andhra Pradesh (Vol. 1-4); *T. Pullaiah et al.*
Flora of the Sawantwadi (Vol. 1-2); *S.M. Almeida*
Flora of Western Tribal Madhya Pradesh; *S. Samvatsar*
Flora of Upper Liddar Valleys of Kashmir Himalaya; *B.M. Sharma & P.S. Jamwal*
Flowering Plants of Thrissur Forest; *N. Sasidharan & V.V. Sivarajan*
Forage Grasses and Legumes; *L.N. Singh, A. Singh & J. Singh*
Grasses and Bamboos of India (Vol. 1-2); *Sunanda Moulik*
Identification of Trees and Shrubs (2nd Ed.); *F.K. Makins*
Integrated Management of Plant Resources; *M.K. Rai & S.K. Verma*
Isthmian Ethnobotany Dictionary (3rd Ed.); *J.A. Duke*
Leafy and Edible Plants of North-East India; *Sudhir Kumar*
Materia Medica (Reprint); *H.G. Greenish*
A Manual of Ethnobotany (2nd Ed.); *S.K. Jain (Ed.)*
The Medicinal and Poisonous Legumes of India; *J.F. Caius*
Medicinal and Poisonous Plants of India; *J.F. Caius*
Medicinal Plants of Bombay Presidency; *S.P. Agharkar*
Medicinal Plants of North-East India; *Sudhir Kumar*
Monograph on Indian *Leucas* R. Br. (Dronpushpi) Lamiaceae; *V. Singh*
Monograph on Indian Subtribe Cassiinae (Caesalpiniaceae); *V. Singh*
The Poisonous Plants of Bombay; *K.R. Kirtikar*
Systematic Studies in Polygonaceae of Kashmir Himalaya; *A.H. Munshi & Javeid*
The Family Euphorbiaceae of Andaman and Nicobar Islands; *T. Chakraborty*
Tropical Planting and Gardening (6th Ed.); *H.S. Barlow*
Vegetable Gums and Resins; *F.N. Howes*
Weed Flora of Kashmir Valley; *M.K. Kaul*

THE POISONOUS PLANTS OF BOMBAY

K.R. Kirtikar
Surgeon, Major

SCIENTIFIC PUBLISHERS (INDIA)
P.O. BOX 91 JODHPUR

The work has been published in the Journal of the Bombay Natural History Society as below:

Vol./Issue	Year of Pub.	Page numbers	Part No. of series	Page number in present book
7(1)	1892	61-76	I	1-15
7(2)	1892	203-207	II	17-21
7(3)	1892	312-317	III	22-27
7(4)	1892	487-493	IV	29-35
8(1)	1893	99-106	V	37-44
8(2)	1893	223-230	VI	45-52
8(3)	1893	331-334	VII	53-56
8(4)	1893	453-461	VIII	57-65
9(1)	1894	42-60	IX	66-84
9(2)	1894	147-176	X	85-113
9(3)	1894	235-258	XI	115-137
9(4)	1894	345-365	XII	139-159
10(1)	1895	88-107	XIII	160-179
10(2)	1895	260-279	XIV	180-199
10(3)	1895	482-502	XV	200-220
10(4)	1895	618-627	XVI	223-231
11(1)	1896	252-261	XVII	232-241
11(2)	1896	606-623	XVIII	242-259
14(1)	1901	20-45	XIX	260-284
15(1)	1904	56-69	XX	286-299

Published by:
Scientific Publishers (India)
5-A, New Pali Road, P.O. Box 91
JODHPUR - 342 001
Tel.: +91-291-2433323 Fax: +91-291-2512580
E-mail: scienti@sancharnet.in
website: www.scientificpub.com

First published in Book form, 2003

ISBN: 81-7233-344-7

Lasertype set: Rajesh Ojha
Printed in India

PUBLISHER'S NOTE

A renowned worker on Indian medicinal plants, Dr. K.R. Kirtikar of Indian Military Service also worked on poisonous plants and his findings were published in a series of papers I-XX in Journal of the Bombay Natural History Society under the title "The Poisonous plants of Bombay" during 1892-1904.

The early volumes of the Journal of Bombay Natural History Society are rare and not available in Libraries. Therefore, we have compiled these papers and presented in a book form for easy reference. Colour plates published by Dr. Kirtikar have not been included due to brittleness of the paper and a few being not available, probably damaged due to brittleness. Figures of these plants are available in many botanical books.

This compilation is very useful to persons of medical profession, pharmacologists, and botanists working on medicinal and poisonous plants.

Although this is not a facsmile reprint of papers yet we have tried our best to publish it like a facsimile.

JODHPUR
8 May, 2003

PAWAN KUMAR

CONTENTS

Reprinted from
Journal Bombay Natural History Society
Vol. 7 pp. 61-76. 1892

THE POISONOUS PLANTS OF BOMBAY

BY

SURGEON - MAJOR K.R. KIRTIKAR, I.M.S.

(Read before the Bombay Natural History Society on 5th April, 1982)

WHEN four years ago Brigade-Surgeon Lieutenant-Colonel I.B. Lyon did me the honour of asking me to supply him with a few concise notes on the Botanical Characters of Indian Poisonous Plants, which he has embodied in one of the Miscellaneous Appendices of his deservedly successful work on Indian Medical Jurisprudence, I thought if an independent series of accurate, well drawn, and properly coloured illustrations were placed into the hands of persons interested in the study of the poisonous plants of Bombay, it would.be a great help towards the ready identification of such plants. Such a series would serve as a companion to Dr. Lyon's work, which is now extensively read all over the Presidency, without attempting to rival it or mar its usefulness. It would, moreover, I thought, enable me to add a few useful hints regarding the plants, which I could not then do from the nature of the work assigned to me by Dr. Lyon and from the necessarily limited space placed by him at my disposal.

Having been engaged for some years past in getting up the illustrations of some of the most typical and usefyl forest and garden plants commonly seen in and around Bombay and Thana, and happening to have in my possession the illustrations of some of our poisonous plants drawn at my request and under my personal supervision by Mr. Isaac Benjamin, I broached the subject of publishing some of them in the Society's Journal to our energetic Secretary, and placed at his disposal my illustrations, offering at the same time to write the letterpress. Mr. Phipson, with the promptness which marks everything he does, whether in connection with the Natural History Society or any other Institution, at once accepted my offer, and promised to supplement my illustrations with a few more drawn by Mr. Benjamin expressly for this series under my supervision. My original pictures are all of the natural size. Most of them, therefore, have to appear as reduced copies of the original to suit the size of the Journal. In each case, however, at the foot of each illustration a note will be added as to what extent the natural size has been reduced. Of Mr. Benjamin's work, I can safely say that, apart from their artistic finish, the specimens depicted are accurate and can be depended upon for the details of their botanical characteristics. They are in every instance copied from nature in their fresh condition, and in each case every attempt is made to secure a typical specimen as far as available.

"What is a poisonous plant ?" — it will be asked. It is as difficult for a Botanist to answer this question as it is for a Medico-Jurist to define"a poison" in works on Medical Jurisprudence. Not even does the Indian Penal Code attempt to define "a poison," be it of vegetable, animal, mineral, or any other origin. Beck, one of the earliest of the standard writers on Medical Jurisprudence, quotes the definition given by Fodere , which, as the former rightly observes, is probably as unexceptionable as any that has yet been attempted. It runs thus — "Poisons are substances which are known by physicians as capable of altering or destroying in a majority of cases some or all the functions necessary to life." This brief definition may be further illustrated in the words of Dr. Francis Ogston so as to restrict the term to "such substances as when exhibited in certain quantities to healthy and ordinarily constituted individuals are capable of producing injurious or fatal effects in a more or less direct and certain way, unless when specially and specifically counteracted." Plants exhibiting such qualities may be looked upon as poisonous. The poison or noxious element may consist of an alkaloid or active principle and may exist in any or all different parts of the organs of nutrition, viz., — root, stem and leaves or their appendages, such as hairs, glands, &c., or in the different parts of the organs of reproduction, viz., — Flower, fruit and seeds. Poisonous plants are more or less speedy in their action, but they may not affect all alike or with equal severity. Their effects vary in an individual under different circumstances. Thus, for instance, the empty or loaded condition of the stomach materially modifies the injurious effects of a poison. The latter state even annihilates the toxic or irritant effects of some poisonous plants. Habit, again, manifestly affects the deleterious effects of poisons. The ganjah-smokers or bhang-drinkers, who respectively indulge in their pipe of the flowering tops of *Cannabis indica* (Indian hemp) or a cold infusion of its leaves, may not suffer anything more than a mere temporary excitement or inebriation, which may pass off without remedy. The same quantity, however, may in the novice or uninitiate produce double-vision, profound narcosis and even death by coma. This illustrates the popular adage that "what is one man's food may be another's poison." This will also explain why I have in the present series tried to illustrate plants and include them among the poisonous, such as have not hitherto been included in the noxious category, nor, indeed, are even suspected as being possessed of deleterious properties. "Forewarned is forearmed." In describing poisonous plants, therefore, it will be my endeavour to embody in this series, not only such plants as have been reputed poisonous from time immemorial, but also those which, within my experience, have struck me as having proved deleterious sometimes to some individuals, although used harmlessly by others.

To District Officers, and particularly to those on whom devolve the magisterial duties of trying cases of clandestine poisoning, and to Medical Officers on whom lies the sole responsibility of identifying and naming the poisonous plants, and thus occasionally helping in the cause of the administration of justice, it is to be hoped that these illustrations may be of some use in a country the vegetation of which is essentially different from that of the land of their birth and education.

I am conscious that the illustrations fall far short of what they might be, though, as I have already said, every attempt has been made to secure accuracy. What merit or artistic beauty they possess is entirely due to the facile pencil of Mr. Benjamin, who has been my valued collaborateur in my illustrations of the Bombay Flora, Phanerogamic and Cryptogamic, and whose eye to details is as trustworthy as it is capable of delineating a charming copy true to nature.

No attempt is made to describe the plants according to their natural orders. Nor indeed is there any order as regards their appearance in this Journal. The plants are depicted just as I came across them, regardless of their virulence or severity of action on the human frame. But each plant as it appears here will be accompanied with a letterpress giving a detailed description of the plant to enable the reader to identify it.

It may perhaps be useful to explain how vegetable poisons act. Their action is either local or remote. When they act locally as, for instance, on the skin or stomach, they irritate the parts *direct*; they are therefore called *Irritants*. The irritation may or may not be followed by inflammation. Their action in this respect is sometimes purely mechanical as, for instance, in the irritation produced by the rigid brown hairs covering the sigmoid rods of *Mucuna pruriens* (cowhage). When poisons act remotely they do so either through the nervous system and are hence called *Neurotic*; or through the heart and are hence called *Cardiac*. Cardiac poisons distinctly affect the heart in the first instance, and cause death by a sudden or gradual failure of its action. Neurotic poisons either affect the brain or the spinal cord singly, or both together. In the first case the poisons are known as *Cerebral*, and produce delirium or torpor, which goes, under the name of Narcotism, the poisons themselves being termed *Narcotics*; in the second case, where the poisons affect the spinal cord, they are called *Spinal poisons*. They cause increase or decrease or total loss of sensation or motion in parts supplied by the nerves issuing from the spinal cord; thirdly, when the neurotic poisons act on the brain and spinal cord jointly they are called *Cerebro-spinal poisons*. In them we find a combination of the symptoms of both the cerebral and spinal poisons. I am not aware of any vegetable poison acting as what is called a *Septic* poison, i.e., producing death by destroying the vitality of blood as is the case in colubrine or viperine poisoning.

STROBILANTHES CALLOSUS — (*Nees.*)
MARATHI — KARAVI कारवी
(Natural Order — ACANTHACEAE)

DESCRIPTION — A shrub 6 to 8 feet.

ROOT. — Bearing buds of the future plant, which are thickly covered over with 8-10 stiff, tough imbricated scales, studded with fine white wavy woolly hairs from three to four lines in length.

STEM. — Erect, caespitose, irregularly quadrilateral, rounded off at the angles; grooved often deeply, throughout, thus marking off each of the four angles of the stem; distinctly jointed like the bamboo down to the central pith; joints bilaterally swollen above the point of juncture, varying in length from a span to a span and a half. Surface of stem irregularly verrucose or "scabrous tubercled," as Clarke calls it, glabrous where there are no warts. The whole stem tapers gradually from the root to the growing tip, varying in thickness at the root from 1/2 to 1 inch in diameter. This tapering is highly characteristics of this plant. The inner substance of the stem throughout consists of close-packed white pith which occupies over three- fourths of its diameter. The pith has an aromatic sweetish smell, not unlike the smell of some of the pith containing stems of the *Graminaceae*. The woody portion is barely a line or two in width or thickness. The outer-bark peels off very thin on scratching, is translucent and of buff colour in the old parts of the plant, purplish or pink at the growing end. The warts do not extend to the inner bark. The inner bark, green.

BRANCHES. — Generally absent, but when given off, always arise at the joints; are seldom subdivided; bearing a pair of opposite leaves; not thicker than a goose-quill; partaking of the quadrilateral nature of the stem; verrucose; buff coloured when old; pink or purplish and downy when young and growing.

LEAVES. — Decussate, arising at the joints only; 7-10 inches long, 3-5 inches broad; crenate; scabrous and ciliated on the upper and under surfaces; "Elliptic cuspidate" — running down into a thin long petiole, 2-3 inches long; nerves well marked especially on the under surface, whitish, and here and there covered with small warts, 8-20 pair. Odour faintly aromatic when the leaf is bruised, some say strongly aromatic. Feel sticky when the leaves are young, from the fluid contained in the hair-cells.

SPIKES. — Axillary single or in pair, bilateral; 1-4 inches long, strobiliform. It is on account of this characteristic strobiliform arrangement of the bracts that the genus derives its name *Strobilanthes*. Inflorescence densely cymose. Peduncles 2-3 lines in length. Bracts $\frac{1}{2}$-1 inch long, concave, glabrous, with entire margins; the lower ones elliptical or oval, tough, green, remote from each other; the lower ones tapering or orbicular, tender, beautifully pink, closer packed, imbricate. The colour of flowers, beautifully purplish -blue, tinged with pink or rosy hue. Dalzell describes the flower as being "deep blue". Nees calls it "*caerulea*." (*Vide* Wallich's Plant. Asiat Rarior. Part III., p. 85). It may be noted that the colour of my illustration is copied from nature, from a specimen growing in my own garden at Thana when in full bloom and fresh. Hooker mentions *Strobilanthes purpurea* as growing in Ceylon (*Vide* p. 183, DeCandolle's Prodromus, P. XI., under *Strobilanthes asperrimus.*)

CALYX. — Persistent, 1/2 inch long; in fruit often exceeding even an inch; deeply quinquipartite; segments or lobes often free down to the base, oblong or acuminate, tough, covered with soft hair, greenish or yellowish white. Segments become tougher and stiffer as they grow older.

COROLLA. — Delicate 1 1/2 to 2 inches long, subequally five-lobed; glabrous without; very hairy-softly so-within.

STAMENS. — Didynamous inserted on the corolla; filaments hairy downwards; auther-lobes bright yellow, two-celled; included.

PISTIL. — Glabrous, elevated on a scarlet or deep-yellow conspicuous globular disc of the size of a millet seed; style filiform, white, slightly bifid, if at all divided.

FRUIT. — A capsule two-celled, 1 inch long, 1/2 inch broad; coriaceous; obovate; loculicidal; valves elastically recurved, carrying the seeds on each half of the capsule.

SEEDS. — Three or four in each half of the capsule; 1/3 inch long, roundish or ovoid, covered over with fine soft down.

TESTA. — Membranaceous.

The tree flowers from June to September. The specimen I have described threw out flowers in my garden, even when very young, from July to September. The idea that this plant flowers only once in seven years is a rural myth. The stem of the plant is an article of domestic economy in rural and even town life. The stem of the plant is cut down when about 6-8 feet long for making the mud- plastered walls of our rural and town huts and homes. It takes each stem about three years to grow to that height, and in the forest it does not flower until fairly grown. The impression, therefore, prevails that it only flowers once in seven years. The fact is when once the plant has flowered, *i.e.*, has grown sufficiently big to be cut down for economic purposes, it is cut down and sold in the bazaars as an article of commerce. The result is, until another plant grows of sufficiently large size in the same place from the same root, for mark you, the stem is caespitose, there are no flowers in the place for a year or two or three which is the time necessary for the 6-8 ft. growth of the plant. As I have said above, the plant has flowered in my garden in the very first year of its growth.

REMARKS — The plant I have described is known on this side of India — in the Konkan — as *Karavi* (Marathi-). In Thana it grows abundantly on the hills dividing the Thana valley from the Vehar and Tulsi valleys. My description is from the outgrowth of the species common here. There is some confusion, unfortunately, but I believe not unnecessarily with regard to the scientific name that should be given to the plant which goes under the Marathi popular name of *Karavi*. The Hon'ble Mr. Justice Birdwood designates Karvi as *Strobilanthes asperrimus* in his Catalogue of Mahableshwar and Matheran plants. Brigade- Surgeon Dymock — the Prince of our living Bombay, aye, Indian Botanists — calls it *Strobilanthes ciliatus*. Colonel Beddome, another well-known name in Indian Botany, designates Karvi as *Strobilanthes grahamianus*.

GRAHAMIANUS

Now when three such eminent authorities-apparently widely differing from each other-writers on Indian Botany well versed in the local Flora they have respectively studied and mastered, designate Karvi in their own special way, while I, on the other hand, follow a defunct Professor from a distant Academy, some apology, some words of explanation, may be deemed necessary, and here I will tender them in all humility.

The term *Karvi* to my native Indian mind is essentially expressive of *economy*. Every plant, therefore, of the genus *Strobilanthes*, whether it be *Asperrimus, Ciliatus, Grahamianus* or *Callosus*, represents a species that is known among the natives of the soil as an economic plant, fit to build up their mud-plastered huts, and as such is known as *Karvi*. What matters it then whether a Bird-wood calls it *S. asperriamus*, a Dymock calls it *S. ciliatus*, a Beddome calls it *S. Grahamianus*, or I, following Nees, call it *S. callosus*? It is *Karvi* after all — a *Strobilanthes* and nothing more. It would still appear to be necessary that I should note the distinguishing points which characterize the specimens which have been named or described by the distinguished and accurately observant Botanists I have named-appearing as they do to differ so much, though in my estimation really not differing to any material extent.

The specimen of the *species*. I have examined is perhaps different from the specimen the Hon'ble Mr. Justice Birdwood, or Brigade-Surgeon Dymock, or Colonel Beddome may have examined, but it is generally and generically speaking *Karvi* or *Strobilanthes* all the same. Specific differences may exist in all of them if each one is compared with the other, and what I have chiefly to point out in this contribution is that if the leaves of the *Karvi* that I describe as *Strobilanthes callosus.* have any irritant properties or poisonous effects on the human stomach on account of its hairs or glandular appendages, Birdwood's *Strobilanthes asperrimus* Karwi, Dymock's *Strobilanthes ciliatus* Karvi, and Beddome's *Strobilanthes grahamianus* may, if similarly carelessly used, produce similar effects on the human body. Let me therefore prepare my reader to note how the plant I have described differs in its minor points from that described or named by Mr. Birdwood, Dr. Dymock and Col. Beddome, respectively. Mr. Birdwood's species has hirsute joints and trichotomous petioles. The species I describe has none of these. Dr. Dymock's description of *S. ciliatus* is very much the same as that of my *S. callosus*. But the description of *S. ciliatus* given by Nees in Wallich (Pt. III., Plant, Asiat. Rar., p. 85) whom Dr. Dymock cites as his authority, is slightly different from the description given by Dr. Dymock. The plant described by Professor Nees has "*Rami supra genicula fibroso-fimbriati*," whereas Dr. Dymock's species is "branchless." (*Vide* 2nd ed., Mat. Med. of Western India, p. 592). The flowers in Nees, specimen are described as "longitudine bractae" "Corolla lutea (?)" — the query Nees own, thus showing that Nees was doubtful as to whether the colour of his species was really yellow. Whereas Dymock is positive about the colour of his *S. ciliatus* being not only blue but *bright blue*.

Leaving these three Botanists aside when I pursue this subject still further and take up Hooker's standard work on the Flora of British India, another question crops up. It is this, I must humbly admit, I find some difficulty-no small one-in following the attempt made by Clarke in amalgamating Dalzell's *Strobilanthes grahamianus* with *Strobilanthes callosus* (Nees). It must be apparent to every reader of Dalzell that his *Strobilanthes grahamianus* is essentially different from *S. callosus.* For *S. grahamianus* has peltate hairs, whereas the hairs of *S. callosus* are longitudinal and tapering at the free end in a sharp point, and are composed of 2-3 longitudinal cells placed end on end. This latter I have verified by Personal microscopic examination. Again Dalzell's *S. grahamianus* has trifid peduncles; the peduncles of *S. callosus* are solitary. *S. grahamianus* and *S. callosus* must therefore on close scrutiny appear to be two distinct species.

THE POISONOUS NATURE OF THE PLANT

With regard to the poisonous nature of the plant, the announcement now made for the first time that the *plant is poisonous*, will, to not a few, if not to all, come with some surprise. But it need not be so, although in the large number of the genera of the natural order *Acanthaceae* there are only a few plants which are known to any irritant or, broadly speaking, poisonous properties. The poisonous quality is exhibited either in the irritant juice of its leaves or in the irritating action of its surface-hairs. Thus, for instance, the leaves of the well- known *Gajakarni Rhinacanthus communis* are well known for the local irritation-often amounting to vesication-they cause in the treatment of ringworm. It is a favorite native remedy for rignworm in certain stages. The leaves of *Blepharis edulis* similarly, which are armed with prickles, — (only an advanced stage of the hairs on the leaves of our *S. callosus*) and the stem of which is still more prickly, Dr. Dymock says "cause redness, burning and itching" (*vide* p. 593, 2nd ed., Mat. Med. of Western India), when they come in contact with the body. Similarly I have found that the pounded or bruised leaves of *Strobilanthes callosus* when taken internally as a cold infusion have sometimes caused irritation of the stomach and produced severe vomiting followed by symptoms of gastritis. The inflammation of the mucous membrane of the stomach thus produced appears to me to be due to the mechanical irritation caused by the hairs on the leaves. Any careless use therefore of the pounded leaves for medicinal purposes is fraught with danger to the mucous coat of the stomach. The infusion should be strained through fine muslin or flannel to get rid of the hairs before use. The fresh leaves are occasionally used by the rural classes as a general tonic, antifebrile and antiperiodic remedy in malarial fevers, and also as a stomachic stimulant and purgative. These remedial uses have the sanction of Nighanta Ratnakara, which serves as a guide to many a native medical practitioner. (*Vide* Vol. III., Arka-Prakash, p. 24, Bombay edition, 1864). It was when the fresh leaves were used for one or other of these remedial purposes that distinct symptoms of gastric irritation showed themselves, for the relief of which latter the cases came under my observation. An examination of the vomits under the microscope showed on each occasion numerous hairs embedded in mucus. It is when the leaves are quite fresh and the minute hairs stiff and erect with the liquid contents of their cells intact, that their irritative potency is most active. Beyond local irritation therefore limited to one organ only, I do not suppose the plant has any other poisonous action. The amount of irritation that its leaves produce when taken in an unguarded way will, I hope, justify my including this plant among the poisonous plants of Bombay.

TRICHOSANTHES PALMATA. — (*Roxb.*)

Marathi — "***Koundal***" (कौंडल)

(Natural order — CUCURBITACEAE)

DESCRIPTION — A large climbing perennial herb; sometimes twisting spirally to a marked degree; a native of forests and field hedges, running over the highest trees at times; distinctly dioecious.

ROOT. — Inclined to be wavy or contorted.

STEM. — Angular or irreguarly rounded; deeply fissured longitudinally; often as thick as a man's arm, says Dr. Dymock, and marked with parallel rows of small irregular warts on either side of each fissure; noduled and jointed; each joint situated at the distance of from $1\frac{1}{2}$ to 2 or 3 inches; giving off leaves or branches at the joints only. The transverse section of a mature stem shows that the longitudinal fissures correspond to the medullary rays and include between each pair of them wedge-shaped woody and vascular bundles, studded with round or oval intercellular canals of pretty large size sufficient to admit an ordinary pin. Dymock calculates these wedge-shaped portions as seven, but I have specimens before me in which they are as many as ten. When in spring or before the rains, the plant is resting and leafless, these intercellular spaces contain air. But when the plant revives, and resumes its activity in the rains, they contain sap which continues to flow through them for some time after the rains.

THE OUTER BARK. — Is light grey or brown, warty, corky, often presenting the appearance of crocodile skin; peeling off easily in irregular bits. *Mesophloem* deep green.

BRANCHES. — Partaking of the spiral or winding nature of the stem; minutely tuberculated; young branches full of greenish pith, Older branches contain brown pith which loses its spongy nature and hardens into a rough friable substance.

TENDRILS. — Three-cleft, oftener bifid, minutely spiral.

LEAVES. — Generally palmate, bright green, membranous; 4-8 inches long, 2-6 inches broad, 3-5 or even 7-lobed; cordate at base; scabrous upon the upper and under surfaces; upper surface more markedly scabrous, and spotted with larger hairs seated on raised circular discs, giving the leaf a characteristic appearance under the magnifying glass. Nerves 3-5 markedly scabrous on the under surface, having a gland or two on each nerve deeply seated. Lobes broad, sometimes entire, sometimes again lobed, and with segments narrow, linear-lanceolate. Margin slightly serrate or dentate; Lobe-divisions sometimes deep.

Petioles also having a tendency to be winding or twisted 1-2 inches long; channelled; several large glands at the apex of petiole.

Stipules single, small, axillary.

FLOWERS. — White, delicate in the female, stout in the male, fading soon after opening over-night or early morning. The plant blossoms during hot and rainy seasons, says Roxburgh, but I have seen male flowers in November and December in Thana in the Judge's garden. This Thana plant is leafless now (March), and will continue so until the next monsoon.

MALE FLOWERS. — Racemed, large white, most delicately fringed with long white branched bifid or trifid filaments. *Racemes* axillary, longer than the leaves, solitary, with a smaller few - flowered second *peduncles* sometimes paired, stout, 5-6 inches long. *Bracts* of the male racemes large, foliaceous, sheathing the flower from a broad base, many times larger than the very short pedicels; ovate; fringed; viscid; covered on the outside with dark green glandular spots of the size of linseed or millet seed. *Calyx* 1- 1-1/2 inch, bract-like; segments ovate, deeply toothed or serrate; tomentose. *Corolla* 4 inches in diameter, hypercrateriform, having the appearance of a flat open parasol with its fimbriae hanging down in beautiful tapers. Petals marked yellow at the base, cuneate. *Filaments* triadelphous. Anthers syngenesious, very anfractuous.

FEMALE FLOWERS. — Solitary, smaller and more delicately fimbriated than the male; axillary; peduncle not so stout as in the male. Calyx-teeth of the female flower less marked. Calyx tube in female short. Petals according to some nearly destitute of fimbriae. *Corolla* smaller than that of the male.

FRUIT. — 2-4 inches in diameter, globose, smooth, of the size of an ordinary orange. When ripe of a bright deep red colour replete with a dirty looking dark greenish pulp in which the seeds nestle.

SEEDS. — Numerous, oblong, compressed, irregularly triangular, obtuse-margined; $\frac{1}{3}$ inch long; according to Dr. Dymock $\frac{7}{16}$ inch long, covered with a blackish shell, and containing a sweet oily kernel.

REMARKS. — With regard to the height of this plant, Clarke (in Hooker's Flora of British India, Vol. II, p. 607) puts it down as "often 30 ft." But I think Roxburgh described it more accurately when he mentioned it as a "native of forests where it runs over the highest trees." The plant may be seen trailing over hedges and branches of trees over several yards. Roxburgh designates the plant monoecious. Wight and Arnott describe the female flower as "*solitary in the same axil as the male*" or occasionally racemose. I have not seen the male and female flowers in the same axil yet; nor on the same plant. But I should be afraid of a definite opinion in the face of such weighty authorities : I would leave other observers to note this point. "Leaves," says Wight, "are *glabrous, sometimes slightly scabrous*." The whole plant, I think, is scabrous to a more or less degree, except the fruit. Observe Clarke's remark at p. 607, Vol. II. Hooker's Flora of British India —"A *Trichosanthes* collected in Margin by Griffith has the leaves with short hairs beneath." Variety *Trichosanthes Tomentosa* is also tomentose beneath (*Heyne in Herb Rottle*).

Clarke describes the fruit as marked with ten orange streaks. I cannot help thinking that when he wrote this he had the *Cucumis Trigonus, var. pubescens* or "*Takmak*" (Marathi) before his mind's eye. I have not seen a fruit of this plant so definitely streaked. Hooker follows this description in the letterpress accompanying Tab. 6873 in Vol. XXII of the 3rd series of Curtis, Bot. Magazine, May 1st, 1886. The plate is a good illustration of the male flowering plant. My plate, be it observed in passing, is of the female plant with the fruit changing colour in the course of its maturity. The colour is rich, but not uniform; it changes from day to day from the bright green of its younger days to the golden yellow, orange, or bright vermilion of advance life, interspersed with all the shades between, often exhibited on one and the same individual fruit, at one and the same time in its later life.

Hooker describes the flowers as sweet-scented. I think to some the odour may prove not particularly agreeable-mawkish if anything. The fruit is seldom if ever pyriform, a form which has been observed by some writers. Dr. Dymock says the number of seeds ranges from 60-100. This is rather a high average for the Konkan, where perhaps the fruit is smaller. 20-30 is as good an average as I can strike. The whole fruit of the plant — the rind and pulp included - is intensely bitter, but, as has been noted above, the kernel of the seed is sweet. Dr. Dymock says the vine is *not bitter* (*vide* "Pharmacographia Indica," Vol. II., p. 71). I have now a section of the stem of the male plant before me. The bark and the wood are both bitter, but not half so bitter as the fruit.

In passing I may here observe that the size of the fruit, as growing in India and as noted by all Indian botanists, is much smaller than that of the same fruit growing in Australia. Baron Sir Ferb. Von Mueller, the accomplished veteran Government Botanist of Victoria, notes that the *Trichosanthes Palmata* growing near Burnett River bears fruit 3-6 inches long, 2-3 inches broad (p. 187, Vol. VI., Fragmenta Phytographiae Australia).

THE POISONOUS PROPERTIES. — The poisonous properties of the plant exist in the pulp and fruit-shell. The pulp acts like a drastic purgative when taken internally as a mere laxative. In producing this effect the plant partakes of the drastic properties of its congener *Ecbalium Elaterium* (Syn. *Ecbalium officinarum*) commonly called the squirting cucumber, which furnishes the British Pharmacopoeia with one of its most powerful purgatives; known as Elaterium, which is the dried sediment of the expressed juice of the fruit. It may be noted that these purgative properties of the respective plants do not suffer on drying. The pulp and shell of *Trichosanthes Palmata*, even if dry, retain all their deleterious element intact. They soon soften when moistened with water, and are as potent as fresh fruit. It seems to me that *Trichosanthes Palmata* is more powerful in its action than *Citrullus Colocynthis* — another of its congeners from the *Cucurbitaceae*, which is a recognized purgative in English and Indian Medicines. According to Charak, an ancient standard authority on Indian Materia Medica, *Trichosanthes Palmata* is a blood-purifier and a nervine tonic. The root of it is besides used as a stomachic tonic. It is in the course of the administration of the root of this plant and of the pulp and rind of the fruit that I have seen the poisonous effects which, I think, are sufficient to justify the introduction of this plant among the poisonous plant of Bombay. Roxburgh had heard of its poisonous effects on birds in his day. He notes that "mixed with rice it is used to destroy crows." Dalzell and Gibson note that it is much esteemed in India in diseases of cattle. Dr. Lyon refers to me (*vide* p. 199, Medical Jurisprudence for India, 1889) as having informed him of the fact that the fruit pulp is used by forest frequenters as a cattle poison. I have since met with cases wherein it has acted as a poison on men. When it is used for poisoning cattle it is mixed up with fodder. Sometimes, however, it has unexpectedly acted as a poison when administered medicinally to cattle for the cure of inflammation of the lungs. The fact that the root is used by Indian villagers for curing acute lung diseases among cattle has long since been noted by Wight. The unsuccessful efforts made at the instance of Sir William O'Shaughnessy and referred to in the Bengal Dispensatory (p. 350) to ascertain whether the fruit had any properties at all, by giving such small doses as three grains thrice daily, need not make us sceptical as to the truly dangerous nature of the fruit-rind and pulp. In small doses the plant root or fruit may act as a stomachic tonic, or may have no sensible effect. But when the pulp of half the fruit, or even a quarter, is used — say if a dram or more in weight — finds its way into the human stomach it does not appear to be free from danger. Administered by ignorant or unsuspecting persons not necessarily with a view to poison, it has done harm; drastic effects have been known to follow. The dry fruit-rind or pulp when smoked is said to act beneficially in the cure of asthma. I have no experience of this myself. On the other hand, I have not heard of any poisonous effects following such administration. But I have known of cases of accidental poisoning when the fruit was administered internally as a laxative and was followed by drastic purgation and irritation of the *primae viae*. I know of no case, however, in which it has caused death, or has ever been used on men for criminal purposes.

Reprinted from
Journal Bombay Natural History Society
Vol. 7 pp. 203-207. 1892

PART II.

(Read before the Bombay Natural History Society on July 4th, 1892)

KAEMPFERIA ROTUNDA - (*Linn., Willd.*)

(Natural Order -SCITAMINEAE).

MARATHI - BHUI-CHAMPA. (भुई चंपा)

Roxburgh has deseribed this species accurately. I have generally followed him and interspersed his description with a few details, whenever found necessary, from Rheede's description of the plant in his *Hortus Malabaricus* (Vol. XI, t. 9, page 17).

A very elegant plant throughout, cultivated in gardens on account of the beauty and fragrance of its flowers. Flowers appear in March and April just before the leaves are thrown out.

ROOT. —Biennial-(I think it is annual); bulbous or tuberous; outside brownish-yellow, covered with a coriaceous memberane; inside yellowish-white; dense; juicy; with numberous white rootlets two or three inches long; bearing fascicles of nnumberous oblong bulbs of the thickness of the thumb, verying from an inch to two inches in length; the bulbs are glabrous, inside mucilaginous.

STEM. — Absent.

LEAVES — radical; petioled; oblong, lanceolate; smooth; never exceeding a cubit in length under ordinary cultivation; usually a foot long in good soil; from four to six inches broad. Very prettily coloured underneath - rich purple; green on the ventral surface. *Petioles* sheathing, uniting into what appears a short stem, as in *Curcuma.*

SCAPES. — just sufficient to elevate the flowers above ground, embraced by a few common sheaths of rich greenish-purple colour, shaded with pink.

FLOWERS. — Scapose; from four to six to the scape; very large, of various colours which are all harmoniously blended in one and the same flower from coerulean-white, pink, yellow to deep purple.

BRACTS — two to each flower, surrounding the base of the germ; the inner one has its apex bifid; the exterior or longest is here only about half the length of the calyx.

CALYX. — Rising from the summit of the root; white, one-leaved, membranaceous; as long as the tube of the corolla; somewhat gibbous; apex generally two-toothed, and of a dotted purplish colour.

COROLLA. — Tube long, slender, cylindric, nearly erect; obliquely funnel-shaped towards the mouth. Petals, 6, in two rows of three each. *Exterior* row of petals drooping, linear, white tinged with purple, with margins involute. *Inner* row of petals has two of them longer than the third when the flower fully opens; erect, lanceolar, acute; colour principally coerulean-white with pink or crimson central and marginal streaks. The third petal inferior, deeply divided into two broad obcordate, deflected, pointed lobes of a deep purple colour particularly towards the centre and base. This deep division of the lower petal gives the flower the appearance of a four- petalled organ.

FILAMENT. — Purple arising from the base of the calyx (Rheede); short, erect, broad, inserted on the base of the uppermost two interior divisions of the corolla (Rox.). The corolla is deciduous; calyx thickens, bearing with it seminal capsules.

ANTHER. — Linear and enlarged with an ovate two-forked yellow-coloured, somewhat recurved crest. Rheede calls it a "cornute yellow tongue."

GERM. — Ovate.

STYLE. — Filiform.

STIGMA — Funnel-shaped.

REMARKS. — The plant figured in Curtis, Botanical Magazine, Vol. XXIII., Plate No. 920, p. 920, is by no means one that would give an accurate idea of the vivid colours of the plant, or the profusion of the fasciculated tubers as seen in the Indian specimens. This can be easily accounted for from the fact that picture was taken from a plant grown in the Brompton Botanic Garden, and necessarily reared under the warmth of a stove and possibly in the absence of bright sunshine. It will be useful to note that Rheede has observed-that Jacob Breynius, a learned authority —in his Prodromus Secund. calls this plant *Zedoaria radice rotunda* (i.e., with a round root). Rheede further notes that the whole plant is used externally as an unguent (Poultice? K.R.K.) to wounds, which it is known in Malabar to cure wonderfully. When applied externally it has been supposed to cause "resolution" where there is "coagulation of blood in the body" — meaning, I presume, congestion. It is said "to consume also all purulent material." The juice of roots taken internally is supposed to act as a resolvent of phlegm, of dropsical affections of hands and feet, and of effusions in joints. Such is also the use of this plant on this side of India. Dr. Dymock speaks of it as a popular remedy in mumps (*Galgand*).

The Poisonous Nature of the Plant.

I have noticed instances of profuse salivation and vomiting produced by the internal administration of the juice of the tubers. They are sometimes used by villagers in their fresh condition in throat affections, known under the generic name of *Ghat-sarpa*, to relieve the sense of dryness in inflammations of the pharynx and tonsils. The fresh tubers are pounded or rubbed on a stone with water, — a couple of them , — mixed with water, a small cupful and drunk sediment and all. This mixture carries with it the active principle or alkaloid of the plant and causes in some cases unpleasant symptoms, such as profuse salivation and subsequent vomiting or retching. The active principle is presumably not unlike what is found in the various species known as the "Galangals" and "Zedoarias," which have a more or less "strong, bitter, pungent, camphoraceous taste." The plant I am describing is, as Breynius has determined, a veritable *Zedoaria*, and a representative of the order *Zingiberaceae* which is noted more or less for its sialogogue properties. How is this salivation caused, when the juice of Bhui-Champa is administered by the mouth? The juice may act in one of the two following ways, or both ways simultaneously — (1) It may act through the stomach on the pneumogastric nerve, and produce *emesis* (nausea and vomiting); or (2) it may act locally on the mouth, i.e., on the lingual terminations of the glosso-pharyngeal nerve, or on the lingual branch of the fifth nerve and produce salivation. Sialogogues are either (1) *topical or direct*, or (2) specific, *remote* or indirect, to follow Dr. Lauder Brunton's classification of them . Ginger, the most typical species of the order to which Bhui- Champa belongs, is classed by him as a topical sialogogue. It is a very pungent substance. Bhui-Champa is by no means so pungent, yet it has an unmistakably "bitter, pungent, çamphoraceous taste." And here I am using the words of our careful and experienced clinical observer, Dr. Dymock.* Now the lesser amount of pungency is a mere question of degree; mere pungency, however, may not have much or anything to do with salivation; for in addition to the pungent element which ginger contains there are other crystallizable and non-crystallizable principles in it. For a fuller knowledge of these, the reader may be referred to the valuable researches of Dr. Thresh in the year-Books of Pharmacy for 1879 and 1882 respectively. It may be presumed that Bhui- Champa possesses some of these principles. It is for the future pharmacologist and chemical analyst to determine what they may be. I am now noting only what has been my clinical experience, and not what I can analytically or experimentally account for.

Now to come to another point. Clinically speaking, intense salivation has been known to be a forerunner of vomiting, or is an accompanying or accessory phenomenon. Dr. Lauder Brunton notes that "the nerves which convey stimuli from the stomach and excite salivation which accompanies nausea are contained in the vagus." I have also the testimony of such a careful observer as Brigade-Surgeon Lieutenant-Colonel Wellington Gray, Principal of Grant Medical College, who, when he was Acting Chemical Analyser to Government and Professor of Medical Jurisprudence, thus observed in his official report for 1874-75 —

"The occurrence of salivation mentioned by Colonel (now General Sir Robert) Phayre as having come on when he began to feel the

*Since these lines were written, Dr. Dymock has departed this life, to the great sorrow of the scientific world. Indian Botany has lost an earnest, indefatigable, quiet and unostentations worker at a time when his mature knowledge was being used by him for the advancement of botanical and therapeutical sciences with the stamp of unquestionable authority - K.R.K.

effects of the arsenic he had swallowed, admitted of the very obvious explanation that *it was nothing more than the increased flow of saliva which is the usual accompaniment of nausea,** no matter how it might have been produced." (Trans. Med. and Phys. Soc., Bombay, no. xii, 1876, pp. 215, 216). The topical sialogogues, such as I consider Bhui-Champa to be, produce salivation by stimulating the salivary glands reflexly through the nerve-terminations in the tongue and mucous membrane of the mouth. "The effect produced by topical or reflex sialogogues," Dr. Lauder Brunton observes, "is not the same for each." "Ether and dilute acids produce a thin watery saliva," says he, "but alkalis cause the secretion of a thicker and more viscid saliva : the former appearing to affect chiefly the chorda tympani, and the latter, the sympathetic." In the salivation which was produced by Bhui-Champa the flow was "thin and watery." Can it therefore be said that the action of the drug is on the chorda tympani? Then there is also the effect of the drug on the stomach nerves. It is a subject worthy of investigation for any practical pharmacological experimenter.

Reprinted from
Journal Bombay Natural History Society
Vol. 7 pp. 312-317. 1892

PART III.

(Read before the Bombay Natural History Society on November 29th 1892)

PYTHONIUM WALLICHIANUM — (*Kunth*).

(Natural Order — AROIDEAE)

MARATHI — (शेवले) (SHEVALE)

An annual plant, erect, glabrous; the flower stalk appearing at the end of the hot-weather, just before the outburst of the monsoons; very common in the Thana jungles; leaf-stalk appears in the rains.

TUBER. — Large, placentiform, fleshy, containing plenty of starch; rounded, with a depression on the top where the scape and petiole arise; whence also several cylindrical, fleshy, thickish roots arise, divided into numerous fine filamentous rootlets. Colour, yellowish brown; size, varying from two to six inches or even eight in diameter from side to side; two to four inches from top to bottom Epidermis rough, pitted, inseparable. Tuber-substance on section whitish, firm.

LEAF. — Large, radical solitary, petiolate, glabrous, umbrella-shaped; tripartitely decompound. *Lobes* deeply pinnatifid; secondary segments or leaflets of the tripartite lobes (pinnae) lanceolate, acuminate, four to six inches long, narrow at the base; deep green in colour; the ribs or principal veins running straight from the mid-rib to the margins at equal distances, yellow in colour on the under-surface of the leaflets, and very prominently marked; these principal veins are ultimately reticulate. The outermost lobes are pinnatipartite. The middle lobe slightly pinnate, or sometimes, as in the accompanying plate, a solitary simple leaflet, ovate, acuminate. Margins of the leaflets entire, sinuated. *Petiole*, solid, succulent, cylindrical; shorter than the scape, dividing into three subdivisions, each called by Blume (*Rumphia*) a "rachis", although the term "rachis", according to Lindley, is strictly confined to the divisions of the petiole of the leaves (*sic* of Ferns, as also to the axis of an inflorescence. The so-called "Rachis" of Blume is deeply sulcate, bordered with a decurrent expansion of the lower margins of the lateral leaflets.

SCAPE. — One or two feet long, cylindrical; oftener compressed about the mid-part; $\frac{1}{2}$ –1 inch in diameter, or of the size of the thumb at the apex of the tuber, where it is loosely enfolded by two membraneous greenish-red scales; the scape of the size of the little finger at the most at the top. Colour variegated, with linear streaks of white, purple, pink, green and brown.

SPATHE. — Terminal, solitary, erect, clavate, ten to twelve inches long, thick coriaceous convolute below *cucullate* above (cucullate meaning-upper margins curved inward so as to resemble the point of a slipper or a hood). The spathe as a whole is cymbiform, or boat-shaped, opening about the middle; greenish- yellowish, brown or purplish; dull opaque outside; minutely reticulate; and pinkish or purplish inside.

SPADIX. — Erect, clavate, of the thickness of the middle finger ordinarily, at its broadest part; of the length of the spathe, and completely covered by it; for about an inch from the base bearing the female organs; thence up to about the middle bearing the male organs; for six inches above that densely covered with numerous rough; irregular, spongy, parallel unequal crests or tubercles tinged whitish, yellowish or brown.

ANTHERS. — Sessile [or placed on very short filaments (*Kunth En. Pl. Vol. III., p. 30*], very densely aggregate in four or five stellate fascicles, oblong, fleshy, yellowish, slightly compressed; apex somewhat abrupt; quadrilocular. *Loculi* narrow, tubular; in opposite, pair; dehiscing from the apex (*Wallich). Pollen* composed of large ovate granules.

OVARIES. — Numerous, close-packed, ovoid, hidden in the convoluted base of the spathe; about fifteen in number, arranged in a spiral manner; unilocular, containing an erect, ovate, solitary orthrotropous ovule, enclosed in a soft fleshy indusium.

Style, cylindrical, somewhat incurved; scarcely longer than the ovary.

Stigma, thick, fleshy, somewhat lateral, and looking downward; trilobate;. lobes equal; triangular, acute with tumid margins, closing upon each other in a somewhat valvate manner; two lateral; one superior.

FRUIT. — A *Bacca* or berry changing in colour from green to yellow and subsequently rich scarlet as it matures; monospermous; scarcely seen. *Seed* — unseen; exalbuminous? (K.R.K.)

REMARKS.

Wallich classes this plant synonymous with *Thomsonia Nepalensis* (*vide* Wallich's *Pl. As. Rar. P. I.* 83, T. 99), also Blume (*Rumphia*, I., 150). Kunth refers this plant to the genus *Pythonium*; the plant itself he names after Wallich. The species is a solitary instance of the genus. I am decidedly disposed to retain the name Kunth gives, as it truly represents its natural appearance. The purple marking of the stem at once recalls to mind the mottled skin of the Python. Wallich names the genus "*Thomsonia*," after the celebrated Dr. Anthony Todd Thomson, for a long time Professor of Materia Medica in University College, London, and distinguished for his botanical tastes and for his genuine pharmacological researches in the early days of that Institution. Wallich, however, adds a note, which is well worth remembering, that the genus so named after Dr. Todd Thomson should not be confounded with another genus similarly named, but differently spelt — *Thompsonia*. The latter genus was formed by Mr. Brown and classed as a genus under the Natural Order *Passifloreae*, in honour of Mr. John Vaughan Thompson.

The genus *Pythonium* is allied to *Arum* and *Caladium*, but differs from them in having its spadix entirely covered by the spathe in the base of the spadix being pistilliferous, its middle part being antherbearing, and its apex verrucose; the loculi of the anthers tubiform and vertically dehiscing at the apex; styles manifest; stigma three-lobed, subvalvate. Blume in his *Rumphia* (Vol. I., p. 146) notes that the genus *Pythonium* is intimately allied to the genus *Amorphophallus*, from which it cannot be distinguished except by its quadriporus anthers and uni-ovuled ovary, the ovary of the *Amorphophallus* being, bi-, tri-or quadri-locular. I may also add that the berry of the *Amorphophallus* is monospermous (one-seeded) or oligospermous (with only a few seeds), whereas *Pythonium* is always one-seeded.

The flower-stalks are seldom if ever allowed to remain in our jungles sufficiently long to bear fruit; long before even the stamens and pistils develop, the flower-stalks are cut down by our jungle-haunters as soon as they appear above the ground and sold in the bazaars with a bunch of the fruits of Kakad (*Garuga pinnata*). These stalks are used in cookery. Curries are made of it, finely chopped, mixed with slices of Kakad as an adjunct. The strongle acid quality of this fruit is decidedly destructive of the acrid property of the flower-stalk. The highly acid fruit of Bilimbi (*Averrhoa Bilimbi*) is also used in the Curries of the flower-stalk for a similar purpose. The Oxalate of Potassium which the Bilimbi contains has no doubt the power of destroying the taste and the potency of the acrid juice. Boiling also with a pinch of common salt removes the acridity. When the flower-stalk matures, it emits the offensive odour of a rotting carcass, and attracts swarms of blue-bottles.

THE POISONOUS PROPERTIES OF THE PLANT.

The whole plant is acrid. It has never been used for criminal purposes, but when used for culinary purposes, it has often irritated the mouth, fauces and pharynx to an alarming extent. The late Dr. Vinayak Govind Gidha, L.M. & S., saw, in conjunction with the late Dr. Sakharam Arjun, some cases wherein much irritation of the mouth and throat followed the culinary use of the flower-stalk. The bulb is acrid likewise; but I am not aware of its ever being used for culinary purposes. Nor do I Know of its ever being used as food in famine times when people are hard pressed and eat whatever can keep off hunger, although in the late Dekkan Famine of 1877-78, in the Nasik District, *Teri Alu* (*Colocasia antiquorum*) appears to have been used. It is well known that the English Cuckoo-pint (*Arum maculatum*) of this order is so poisonous that even a small piece of its leaf when eaten by children has been known to produce convulsions. But the chief point I wish to note in connection with it is, that although the root and the plant are highly poisonous, the poison may be removed from them, and a wholesome flour made from the root. — (*Anne Pratt.*) The other representatives of the Arad order, such, for instance, as the plant I am now describing, are equally noted for the readiness with which they part with the acrid element they contain, especially by means of heat. Long boiling before using the flower-stalks renders them harmless, though even sometimes the acridity is perceptible. It is not, therefore, always agreeable to eat curried Shewla, though it may not be always positively unsafe to do so. The fact of the acridity disappearing on boiling is indicative of the volatile nature of the acrid element. Dr. Sakharam Arjun has noted his experience regarding the poisonous properties of the root of *Phthonium Wallichianum* in a paper which, if I remember right, he read some years ago before the Grant Medical College Society. But not having all the numbers of the Society's Transactions and printed Proceedings, I regret I am unable to cite the cases he mentions. I know from my own practice that in several instances severe irritation of the fauces has followed the internal use of the flower-stalk followed by a temporary oedema of the mucous membrane of the uvula and pharynx. In Dr. Sakharam's and Dr. Vinayak Gidha's practice the tongue was noted to have become rapidly swollen, a phenomenon not unknown in the poisonous symptoms following the ingestion of the leaves of the English Cuckoo-pint, a congener of the Bombay Shewla. The curried Shewla, when not acrid, or when the acridity is "just passable", or faint, is by no means an unattractive or unpalatable table delicacy, but you can never know when the throat or the tongue may not have to pay the penalty of even a careful or moderate indulgence in a delicious but treacherous curry.

Reprinted from
Journal Bombay Natural History Society
Vol. 7 pp. 487-493. 1892

PART IV.

TRICHOSANTHES CUCUMERINA (*Linn.*)
MARATHI कडुपडवल; रानपडवल (KUDUPADVAL, RANPADVAL)
(NATURAL ORDER — CUCURBITACEAE)

SYNONYMS :— T. lancinosa.
T. pilosa.
Bryonia umbellata.
Cucumis Missionis.

The plant is a scandent herb, more or less pubescent; annual strictly of the rainy season. Has a faint disagreeable odour. Dioecious; or "less often (? when cultivated) monoecious" — Clarke (*Hooker's Flor. Br. Ind.*)

ROOT — Somewhat tuberous, lying deep in the ground; slightly woody, fibrous, whitish in colour.

STEM — Herbaceous, with a few woody fibres, pentangular, striate watery; having a tendency to the formation of twisted irregular nodules.

LEAVES — 3-6 inches long; 2-4 inches at base; ovate-cordate; slightly lobate; lobes five, not distinctly marked; margin broad and deeply curved at the base, serrate throughout; acuminate. Deep green; faintly cream-coloured here and there; puberulous. *Tendrils* wavy, finely spiral, bifid usually; sometimes trifid or undivided; sometimes displaced by irregular nodes. *Venation* of leaves strikingly whitish; nerves prominent on the under surface. *Petiole*, slender, 2-4 inches long; grooved often deeply; wavy.

FLOWERS. — Buds oblong-rotund. MALE FLOWERS. — Peduncles in pairs; often in racemes with short pedicels at the apex of long peduncles. Size of flowers, small; colour white; petals five, fringed with delicate fibrils; star-shaped, with parallel nerves; crisped and contorted. *Bracts*, large. *Calyx*, 5- sepalled, 1 inch long, tubular; sepals very narrow, green, cuspidate. *Stamens*, 3, short, greenish, almost included, connate; long-linear; 1 one-celled, 2 two-celled; cells conduplicate, yellowish.

FEMALE FLOWERS. — Solitary, often opposite a leaf; calyx and corolla as in male. *Style*, greenish; *Stigma*, trifid and whitish.

FRUIT. — A pepo; 2-4 inches long; ovate, pointed at both ends; with short peduncles which are incurved, dark green and shining, shape oblong-rotund; carpels imperfect; externally marked greenish-white in the younger stage with broken longitudinal lines from apex to base. The colour of the fruit is greenish when first formed, gradually changing into yellow, orange, and deep scarlet as the fruit matures: pulp cream -coloured when the fruit is young.

SEEDS. — $\frac{3}{4}-\frac{1}{2}$ inch, compressed or corrugate; arranged in rows of three, transversely or horizontally; large; varying in number from 20—30 or more, surrounded by a thin foetit bitter watery pulp which assumes a deep red colour when the fruit matures. Shape of seed oblong-plane; angulated at its attachment to the funicle. The fruit matures in September and October. Abundant in the neighbourhood of Bombay.

POISONOUS PROPERTIES.

The plant has a distinctly emetic effect. Rheede (in *Hort. Malab.*, Vol. VII., pp. 107-108) mentions that the juice of the plant produces vomiting. The juice of two ounces of the root acts as a violent drastic purgative, not unlike *Elaterium* which is a product of the squirting cucumber of the same Natural Order. The plant is said to cure quotidian and quartan fevers chiefly by inducing vomiting. The drug, therefore, has to be used with caution. The leaves, root and fruit are all bitter, and act equally powerfully.

GLORIOSA SUPERBA (*Linn.*).
MARATHI — खडयानाग KHADYA NAG
(NATURAL ORDER — LILIACEAE.)

The plant is an annual climber, wild in hedges and jungles; often cultivated in gardens for its beautiful flowers.

ROOT — Tuberous, fleshy, budding from the convexity above; tubers cylindrical or flattened, in some parts sometimes slightly twisted. 4 to 9 inches long; about an inch thick. Substance internally white, mealy, juicy, with a slight acrid odour. The external appearance of the tuber is brown. This is due to the thin paper-like epidermis which encloses it at all parts of the bulb, except the growing point, which latter looks like the eye of a potato. The epidermis is easily removed by the mere handling of the bulb, and is fragile. The old bulb, as the plant matures, shrivels up, and gradually throws out a new bulb at right angles which terminates in a growing point. It is from this growing point that the future plant sprouts out. The rootlets are chiefly confined to the part of the bulb directly opposite to where the stem is situated.

STEM. — Usually a single stem arises from the bulb. It may branch off almost immediately after it leaves the tuber. The stem is put down as 10 — 20 ft. by Hooker (*Flora Br. Ind.*), but is often much longer than that in the jungles, where it is seen in its natural condition. When the stem branches off immediately on leaving the tuber underground, three or four slender cylindrical green sprouts shoot up above the ground, thus making it appear that they are separate plants from one and the same bulb; but it is not so. One bulb, one plant is the rule invariably in a number of bulbs I have carefully examined. The stem is nearly cylindrical, and though herbaceous, is tough and interspersed with woody fibres in close-packed bundles. The medulla of the stem is slightly pithy and full of watery juice when fresh; hollow when the plant dries up and the pith shrivels up. At each insertion of the leaf, which is arranged in pairs in an opposite manner, the stem deviates from the median line, thus giving the whole plant an angular or regular wavy appearance.

LEAVES. — Sessile or short petioled; the inferior ones opposite and oblong; the superior one single, ovate-lanceolate; the other leaf replaced by the flower-stalk, when the plant is in flower. The length of the leaves varies from six to eight inches; the greatest breadth is often as much as two inches. *Venation*, parallel-nerved. Mid-rib prominent, terminating into a stiff spiral coil, which resembles a tendril. This is a marked characteristic of the plant, which, so far as I know, has no parallel in the Vegetable Kingdom.

FLOWERS — Solitary, on long slender variously curved peduncles, 4-6 inches long; deflexed often giving the flowers with their reflex petals the appearance of a slender-winged butterfly poised in the air. It is this that gives the gorgeous flowers their unique appearance, and shows off their rich, varied, and ever-varying hues.

PERIANTH — Of six petals, persistent. *Flower-buds* conical or pyramidal, bright green.

PETALS — 3-4 inches long; linear-lanceolate, narrow, undulate or crisped; bright green with a pale yellow margin when fresh opened; bright yellow up to a third from their insertion, and scarlet up to tip when full-blown; gradually converting into rich scarlet, with golden-yellow margin and mid-rib when the ovary is maturing into fruit and growing in size. The flower retains its beauty for several days. The petals are persistant, and don't fall off even after the plant has dried. They fall off long after the fruit and seeds are mature. It is this that makes it so appreciable as a forest and garden beauty. The petals, which are reflexed when fresh-open, assume a spreading aspect as they grow older.

STAMENS — Six hypogynous; radiating from the ovary.

Filaments — Filiform, sometimes stout; bright green when the flower first opens; gradually changing into bright golden yellow as the flower matures.

Connective — Greenish; changing into golden-yellow.

Anthers — Large, semilunar, versatile, golden-yellow, extrorse; often dorsifixed.

PISTIL : —

OVARY — 3-celled, superior, as is characteristic of the natural order *Liliaceae*; the carpels deeply grooved in the centre, thus giving the fruit the appearance of a six-lobed body, though actually three-lobed, as is clearly shown in figures 4 and 5 in the accompanying plate drawn from nature.

STYLE — Filiform, deflexed almost at right angles from the very apex of the ovary, tapering into a trifid stigma from a broadish base, greenish just as the flowers open, gradually changing into yellow. Sometimes the style remains long after the petals have fallen.

FRUIT — A capsule 2 inches long; an inch broad when full developed; coriaceous; septicidal.

SEEDS — Numerous, subglobose, in dense double rows in each of the three carpels of the capsule, *Testa* spongy white when young, covered with a brilliant rich scarlet mucilaginous coat when the fruit matures.

EMBRYO — Cylindrical.

REMARKS — *Gloriosa superba* is essentially a monsoon plant; it begins to sprout out in the early part of the rainy season, and dies soon after the rains. The bulb remains dormant throughout the cold weather. It flowers and bears fruit from July to September. Should the plant be cultivated as a garden beauty or ornamental trellis-plant, it is best to re-pot or re-plant it in February or March, or at the latest in April, when the terminal bud of the tuber begins to sprout out. Dalsell and Gibson call it "Buchnag". This is a mistake. Buchnag is the root of *Aconitum feror*, which is different in form and appearance. There is no doubt, however,, that the tubers of this plant are adulterated with the tubers of the real Buchnag. Dalzell and Gibson also call the plant "*Kalawee*" This is evidently a misprint, and meant for the Marathi synonym. "Kala-lawee"(कल लावी) Surgeon General Edward Balfour, of Madras, states that the petals of the flower of *Gloriosa superba* are fringed (*Cyclopoedia oj India*, Vol. 1.). It is not so. There is no fringe of any kind on any part of the plant. The colour of the flowers depends a great deal on the amount of the sun or the intensity of it, to which the flowers are exposed as they are maturing and expanding. The specimen from which our draw -ing is made was obtained from a plant which flowered under a persis- tent cloudy sky; the scarlet colour is therefore not quite so deep or bright as it might be, and very often is. Firminger speaks of a caterpillar infesting it. He says that if not looked to, the plant is entirely consumed by the caterpillars. I am not aware of the plant ever being similarly attacked or even approached by devouring insects, on this side of India. It would be useful to have more information on this point.

The Poisonous Properties.

This plant is one of the nine secondary poisons mentioned by Hindu writers. They are given as follows in *Nighanta Ratnakar* (Vol. III., p. 255) —

1. Euphorbia neriifolia.
2. Calotropis gigantea.
3. Gloriosa superba.
4. Abrus precatorius, red and white.
5. Nereum odorum.
6. Strychnos Nux -Vomica.
7. Datura alba and fastuosa.
8. Jatropha Curcas.
9. Papaver somniferum (its product opium).

Very few cases of poisoning are on record, but the plant is well known all over the country as a virulent poison, although it is said to be useful in fevers as an antiperiodic. Dr. Norman Chevers mentions two cases (*Medical Jurisprudence*, p. 284), in one of which the effects are stated to have been those of a Narcotico-irritant poison; it is not stated what poisonous symptoms were observed in the other. In the one $\frac{1}{12}$ tola and in the other 2 tolas of the root caused death. Brigade-Surgeon Lyon mentions (*Medical Jurisprudence*, p. 454) a case reported in the *Indian Medical Gazette*. 1872, p. 153, in which the powdered root (quantity not mentioned) produced the following effects — "Symptoms of poisoning appeared in half an hour, and were :- retching, violent vomiting, spasms and contortions of the body, with fearful racking pain; from time to time there were short intervals of relief, followed by recurrence of the same symptoms. Death took place in four hours. The *Post mortem* appearances were congestion of the brain and its membranes with extravasations of blood. The lungs, liver, and kidneys were all deeply congested. The gastric mucous membrane showed signs of inflammation. The peritoneal covering of the fundus of uterus (unimpregnated) was also found inflamed."

Ancient Hindu writers agree in attributing violent emetic properties to the root; it is also said to cause abortion,. and as such prescribed by Hindu physicians for expelling retained after-births. Dr. Dymock, in citing the researches of Warden, says that there are two resins and a bitter principle in the root. Warden names this bitter principle *superbine*, and considers it identical with that of *Urginea Seilla*, the ordinary Squill. The bitter active principle of Squill, says Dr. Lauder-Brunton, is a glucoside Scillitoxin or Scillain.The Scillitin of some authors is probably slightly impure Scillitoxin (*Pharmacology*, p. 962). Squill is classed by "Lauder- Brunton among stimulants of the Cardiac muscle, when moderate doses are given. When, on the other hand, larger doses are given, it acts as a "Cardiac poison" (p. 276, Op. cit.). In such cases, that is to say, where the dose is large, "the stage of stimulation is followed by one of peristaltic action and final arrest in Systole." It is a pity the case quoted by Dr. Lyon gives no information regarding the *Post mortem* condition of the heart, although the condition of every other organ is noted. It will be useful to note this condition in all future observations with a view to verify the opinion of Warden. In excessive doses, "the operation of Squills," says Dr. Waring (*Therapeutics*, p. 489), "is that of an acro-narcotic poison, 24 grains having proved fatal." Squill is known to cause nausea; a small dose may act with extreme violence. I have already said that the plant is well known among the Hindus as possessing emetic properties. The conjecture of Warden is in my opinion based on a striking similarity of the physiological effects of the two plants on the human body.

Reprinted from
Journal Bombay Natural History Society
Vol. 8 pp. 99-106. 1893

PART V.

CUCUMIS TRIGONUS. — (*Roxb.*)
(Natural Order — CUCURBITACEAE)
MARATHI — KARIT. (कारीट)

The plant grows wild in the Konkan; it is an annual, succulent, scabrous creeper, trailing along the ground or over hedges for several yards at a stretch.

ROOT. — Perennial; trailing a good distance above ground sometimes, throwing out rootlets into the soil; reddish outside, green inside; juicy, bitter.

STEM. — Succulent, quadrangular; tendrils short, wavy, simple lateral and formed of abortive stipules. Rarely bifurcated at apex.

LEAVES. — Alternate; deeply palmately-lobed, cordate at base; lobes 5, rounded, repandly and sharply toothed; margin crenulate; colour deep green. "Lobes of leaves," says Wight, "very broadly ovate and almost touching each other at their broadest part, sinus rounded." Roxburgh, however, states that the lobes are quite distinct, sinus widening upward. These appear to me to be mere differences in development due to local influences. PETIOLES — $\frac{1}{4}$ to $\frac{1}{2}$ inch. The size of leaves C.E. Clarke puts down as 1-2 inches in diameter (Hooker's Flora Br. Ind., Vol. II, p. 619), but I have seen the leaves quite double that size if not more. Leaves and petioles "scabrid, not softly hairy," says Clarke, but the plant is both scabrous and covered with soft hair throughout. The hair markedly white.

FLOWERS. — Monoecious, axillary, unisexual, peduncled.

Male flowers.- Crowded in axils of leaves, opening in succession, 1-1$\frac{1}{2}$ inch in diameter.

Female flowers. — Solitary short-peduncled, 1-1$\frac{1}{2}$ inch in diameter.

CALYX of male and female flowers 5-toothed; tubular -campanulate; segments subulate, scarecely of the length of the tube; calyx of female flowers short.

COROLLA of male and female flowers 5-parted; petals entire, scarcely united together or with the calyx; those of the female flowers being sessile besides. Nerves prominent from base to apex.

STAMENS. — 5, triadelphous.

FILAMENTS. — Inserted below the mouth of the calyx-tube.

ANTHERS. -Exserted from the tube, free; one, 1-celled; two, 2-celled; cells conduplicate; connective produced in a crest.

PISTIL —

STYLE. — Short.

STIGMAS. — 3, very thick, obtuse; 2-lobed, velvety.

OVARY. – Inferior, adhering to the calyx-tube; of 3 carpels, densely hairy; hair short, white, soft-silky. Placentas, 3 parietal.

FRUIT. — A pepo; fleshy internally, pale cream-coloured; coriaceous externally; indehiscent; oval, rounded at both ends or sometimes flattened; obtusely 3-angled, the depressions at the angles corresponding to the union of the carpellary margins; about $1\frac{1}{2}$ long, $1\frac{1}{4}$ broad; often almost twice as large in Thana. The outer surface of the fruit is not *smooth* as stated by Roxburgh, but tuberculate; tubercles of the size of a small pin's head; brown when old; irregularly scattered all over the fruit, but principally over the mid-portion of it. Each tubercle is capped with a fine white soft hair. These tubercles are liable to be detached by the gentlest touch as the fruit gets old; hence they are often not observed when the older specimens of fruit are examined. But yet their position can be distinctly noted on the epicarp by a depression surrounded with raised circular yellow markings. The readiness with which these tubercles get detached may possibly account for Roxburgh's saying that the fruit is smooth. The epicarp is leathery, $\frac{1}{12}$ inch thick; striated; striae 10, irregular, varying from light to deep yellow colour. The fruit keeps for several months without shrinking on account of its tough coriaceous exocarp or rind.

SEED - Sharp, ovate, compressed, not marginated, acute at hilum, mostly smooth, enveloped in a juicy mucilaginous arillus of white colour.

Number varying from 150-250. Even a small fruit, one inch in diameter all round, was found by me to contain as many as 120 seeds.

TESTA. — Coriaceous.

ALBUMEN. — None.

EMBRYO. — Straight

RADICLE. — Next to hilum.

COTYLEDONS. — Foliaceous, palmate-nerved.

REMARKS

The synonyms of this plant have been variously given by various Botanists. What principle they have followed it is difficult to understant. In describing the Tribe *Cucurbiteae* of the Natural Order *Cucurbiteae* — a tribe to which Wight and Arnott have consigned the species we are now describing, viz., *Cucumis trigonus*, they very rightly observe (*vide.* their Prodromus Flora Ind., Vol. I., p. 341), that the difficulties attendant upon the description of the species from dried specimens are insurmountable. They admit, however, and very rightly too, that they consider it "more prudent to follow Roxburgh as closely as possible, as his opportunities of examining both seeds and stamens must have been most ample," as they really were, for his description appear as a general rule drawn up from fresh specimens — a circumstance which must have given him greater facilities for noticing the minuter details of the species he describes. The following are the various synonyms of *C. trigonus* —

C. madraspatanus Roxb. (Hooker's Fl. Br. Ind., Vol. II., p. 619). (Wight and Arnott's Prod. Flor. Ind., Vol. I, p. 342).

C. turbinatus, Roxb. (Hooker op. cit.)

C. melo, var. *agrestis*, Naud. (Hook. op. cit.)

C. pubescens., Willd. (Hook. op. cit.)

C. pseudo-colocynth, Royle. (Hook. op. cit.)

C. eriocarpus, Boiss, et Noe (Naudin Ann des. Sc. Nat., 4 Series, Vol. XI, pp. 1-87); (Hook. op. cit.)

Bryonia cullosa, Herb Rottler (Hook. op. cit.)

C. pyriformis, Roxb. (Naudin op. cit.)

C. villosus, Boiss. et Noe. (Naudin op. cit.)

Why the plant is considered synonymous with *C. madraspatanus* it is difficult to understand, especially as Roxburgh describes the latter as a distinct plant, having the fruit no bigger than a partridge's egg; whereas the size of the fruit of *C. trigonus* Roxburgh puts down as that of a pullet's egg. Moreover the fruit of *C. madraspatanus* is oval, downy, maculated, without any tendency to be three-sided. The fruit of *C. trigonus*, besides, is bitter and not edible, whereas that of *C. madraspatanus* is said by Roxburgh to be much used in food by the natives and much esteemed. The synonym *C. pseudo-colocynth* appears to be used by Dalzell and Gibson on the authority of Royle. The synonyms *C. pubescens* (Willd.), as given by Clarke in Hooker's Flora of British India, is somewhat puzzling. Dalzell and Gibson describe it in their Bombay Flora (p. 103) as a distinct species from *C. trigonus*. and add that it is cultivated in Sindh under the name of "*Chiber*." It is 1- $1\frac{1}{2}$ inch long. This is "confusion worse confounded." The fruit known as *Chibud* or *Chiber*, as Englishmen would ordinarily pronounce it, is strictly speaking the genuine *Cucumis melo* and is a much larger fruit than either *Cucumis pubescens* or *Cucumis trigonus*. It is besides edible uncooked. It is seldom if ever cooked. Again Dalzell and Gibson make *C. madraspatanus* and *C. turbinatus* synonymous with *C. pubescens*, although Roxburgh describes *C. turbinatus* as a distinct species with a turbinate fruit, pyriform in shape, absolutely three-sided, with much larger flowers than *C. trigonus*. Note again another source of confusion. Dr. Lyon calls *C. trigonus*, *Indrayan Bislumbhi*, on the authority of the Pharmacopoeia of India, edited under the supervision of able Indian Botanists. But *Indrayan* is *Citrullus colocynthis*. In this I am borne out by Dr. Dymock; and *Indrayan* is a much more powerfully drastic purgative, with a larger and much rounder fruit, of the size of an ordinary orange or wood-apple. It may be noted, however, that *C. trigonus* is occasionally used as a substitute for Colocynth to produce purgation. To pass on to another writer — Naudin most emphatically says that *C. trigonus*. is certainly the species which M.M. Boissier and Noe have described under the name of *C. eriocarpus*. Naudin adds further that it will be necessary also to unite to this species the one described by Baron Sir Ferdinand von Mueller under the name of *Cucurbita micrantha*, an Australian plant which is probably a species of *Cucumis* and which appears to be analogous to the plant under description (J. Mitra).

A still further confusion arises in the naming of the fruit known in the Konkan as *Takmak*. Dr. Dymock calls it *C. trigonus*, var. *pubescens* (*vide* Dr. Dymock's Materia Medica of Western India, 2nd Edition, p. 339). Dr. Dymock says *Takmak*. is less bitter than *C. trigonus*. As a matter of fact *Takmak* is not bitter at all. If any thing, it is sweetish; if not that, it is almost insipid. It is not therefore easy to find out why *Takmak* is considered to be a variety of *C. trigonus*. The two plants appear to me to be quite distinct. *Takmak* is edible and perfectly harmless. It is cooked as a vegetable. It can be eaten even uncooked, though never much valued in any condition. Wight and Arnott's description of *C. pubescens* shows that the fruit of it is no bigger than about 1 or $1\frac{1}{2}$ inch long; whereas the fruit of *Takmak* is often as large as a medium-sized pomelo or water-melon, 6 inches long and 9 inches sideways; and it is glabrous. The fruit of *C. pubescens*, on the other hand, as described by, Willdenow, and Wight and Arnott is oval and pubescent. It is doubtful therefore whether *Takmak* could be called *C. pubescens*, or a variety of *C. trigonus*

Kurz separates *C. trigonus*. wit *solitary* peduncles from *C. pubescens* with *Clustered* peduncles and makes the latter a variety of *Cucumis melo*, Linn. (*vide* Journal Asiatic Soc., 1877, Part II, page 103). I think Kurz is right in doing so. In every way *Takmak* is more allied to *C. melo* than to *C. trigonus*. The only objection I have to calling *Takmak C. pubescens* is that the size of *C. pubescens* as described, namely, that of a partridge's egg, is considerably smaller than that of the smallest sized *Takmak*. On the whole, it appears necessary to write a fresh description of *Takmak* and rename it once for all. The present name it has been given as a pubescent variety of *C. trigonus*, is morphologically inaccurate and unsatisfactory. The plants marked by Indian Botanists as *C. pubescens* are said to have been reduced to *C. trigonus* by Naudin in the Kew Herbarium. It is here, I believe, that the initial source of inaccuracy lay. For although that illustrious French Botanist can claim the credit of having furnished us with an exhaustive monograph on the Cucurbitaceous order, he has added to the confusion of synonyms in no small degree.

From the above remarks it will be amply apparent that there is a considerable difficulty in confirming and finally establishing the synonyms of *C. trigonus*. It seems to be absolutely necessary to remove all this confusion by a fresh and exhaustive attempt at identifying the plants and renaming them after a careful examination *locally*. of every species belonging to the Cucurbitaceous order. One thing must be remembered particularly, with a view to obtain a thoroughly accurate description. The Cucurbitaceous plants are extremely delicate, especially as regards their hairy appendages. They must therefore be examined *in situ* — in the places where they grow; not certainly from dried specimens, nor even from fresh specimens taken to a distance and examined leisurely, perhaps long after the delicate hairs have withered or falled off, and thus destroyed the minuter distinguishing marks so necessary for accurate identification.

As regards the root of *C. trigonus* Clarke (in Hooker's Flora of British India *loc. cit.*) remarks that from the perennial character of the root of *C. trigonus* alone can it be distinguished from *C. melo* according to Naudin, but says Clarke, "the examples seen and collected in India *are almost invariably less than one year old*." (*N.B.* — The italics are mine. — K.R.K.) As a matter of fact, it is not so. I have seen the plants growing wild in the jungles and hedges of Thana and I have observed the same root throwing out fresh sprouts every rainy season for years together. The plant dries up after the monsoons and the fruit may be gathered even in the hot weather following. The plant is so persistent that the fruit can be seen hanging on the dry creeper even when every other part of the creeper withers and dries up. The fact of the fruit remaining hanging fresh as ever even in the succeeding hot weather, is, I admit, no indication whatever of the living condition of the root underground. But wait till the next rainy season, and you will find a fresh plant from the old root, from where the creeper of the former rainy season sprouted. The root does not dry up, decay or die, even although the surrounding soil is parched and cracked in the hot weather, with hardly any moisture to nourish it. The perennial nature of the root, therefore, must be considered undeniable, even under such adverse circumstances.

With regard to the flowering of *C. trigonus* Rheede observes (Hortus Ind. Mal. Vol. VIII, Page 21) that the plant grows in jungles around Cochen, flowering and fruiting throughout the year. I may observe that in the Konkan, *C. trigonus* flowers and fruits only in the rains. It begins to dry up as soon as the cold weather sets in, and is shrivelled and leafless before the hot weather is on. The fruit, as I have already said, remains hanging unshrivelled and unshrunken even in the succeeding hot weather.

THE POISONOUS PROPERTIES

Being a congener of Colocynth, one would naturally expect that the juice of the bitter fruit of *C. trigonus* would be a purgative of the irritant type, and so it is more or less. Dr. Lyon, however, states in his Medical Jurisprudence (p. 199) that the plant possesses another poisonous property. He says that in 1883 a case was reported to the Bombay Chemical Analyser's office in which it was stated that this plant had been administered for the purpose of procuring abortion.

In the Pharmacographia of India (Vol. II., p. 67), Dymock, Hooper, and Wardell note that they digested the dried fruit with 84 per cent. of alcohol and concentrated the resulting tincture until most of the alcohol had been expelled; they then agitated the mixture with water and petroleum ether. This solution, still containing some alcohol, was heated on the water bath to get rid of the alcohol; the remnant then mixed with water and agitated with acetic ether containing some acetic acid. This yielded a reddish brown extract, very bitter, and partly soluble in boiling water. The insoluble residue was brittle when cold and very bitter, and had the properties of a resin, which would appear to correspond with the resin of Colocynth. The scientific world owes a deep debt of gratitude to this trio of distinguished pharmacologists for thus advancing our knowledge regarding the active principles of *Karit*.

I need hardly enter here into a detailed description of the poisonous properties of Colocynth. They are well known to every student of the British Pharmacopoeia. Suffice it to say that Colocynth and its congeners are irritant drastic purgatives. These properties were well known to the ancients, both Eastern and Western. I may note, however, what has been observed by Dr. Schmeidberg of Strasberg (*vide* Elements of Pharmacology, Dixon, 1887, p. 109) with regard to the relative therapeutic action of the crystalloids and colloids in

Colocynth, and presumably in the allied species like *Karit*, that crystalline soluble Colocynthin, when pure, does not, under all conditions, cause diarrhoea; hence it is apparent that the crystalline element needs the presence of colloidal substances such as occur in Colocynth, for it to be with certainly carried into the intestine to produce any irritant or drastic effects.

It will be perhaps easy to account for the abortifacient effects of *Karit* if we remember how powerful irritants of the intestinal canal and of the rectum especially affect the pelvic organs. Increased peristalsis produced by intestinal irritants known as "aperients" means increased hyperaemia of the mucous membrane of the intestines. This hyperaemia extends to the neighbouring pelvic organs when powerful irritants attack the lowermost part of the intestine. "Thus in gravid uterus," says Schmeidberg, "contractions may be caused ending in abortion and premature labour." It has ever been held a golden rule in pregnancy not to give acrid or powerful purgatives at any time during pregnancy. The abortifacient effect of *Karit* may be presumed, in the cases recorded, to be probably of a secondary nature, although it is not so stated in the Bombay Chemical Analyser's report, there being no details of the case forthcoming in that report.

Reprinted from
Journal Bombay Natural History Society
Vol. 8 pp. 223-230. 1893

PART VI.

TERMINALIA BELLERICA. — (*Roxb.*)
(Natural Order — COMBRETACEAE)
MARATHI — BEHEDA. बेहडा

A very large tree with rusty pubescence on young branchlets and calyx; attains a heigh of 60-100 feet; trunk tall, erect, regularly shaped; branches spreading, forming a coppery tinted bright, broad, massive crown when young; bright green when old. Youngest offshoots beautifully crimson.

STEM. – "Bark $\frac{1}{2}$ inch thick, dark grey, uneven, and tessellated by broad longitudinal furrows crossed by short narrow transverse wrinkles, the old bark exfoliating in dry corky scales. Wood lightly grey, or yellowish, open and coarse grained, easily worked but not durable; no distinct heartwood" (Brandis). The girth of the stem is 6-10 feet, at times 10-20 feet.

LEAVES. — Exstipulate, alternate, crowded about the extremities of branches; crenulate, pubescent; when quite young, of a bright reddish or copper colour, tinged with bright crimson; older leaves bright green, pale beneath; broad elliptic or ovate elliptic, 3-8 inches long, 2-3 inches broad; base often unequal; the lower margin of the leaf tapering as it approaches the petiole, and finally merges into the upper margin of the petiole, leaving the petiole slightly grooved on the ventral aspect. Apex of the leaf obtuse, retuse usually; sometimes accuminate, especially in the larger leaves; margin of the leaf entire; main lateral nerves arcuate, prominent, 5-8 on either side of the midrib; often reddish.

PETIOLE. Roundish, longer than $\frac{1}{3}$ length of the leaf. The tree sheds its leaves from January to March.

FLOWERS. — Small, male and hermaphrodite mixed on solitary simple spikes which are sometimes erect, sometimes bent, sometimes drooping; spikes 3-6 inches long, arising from the axils of fresh leaves just before or about the same time as the year's tender leaves shoot out. Sometimes the spikes shoot out from below the insertion of the fresh leaf; not unfrequently from either above or below the cicatrix at the insertion of the previous year's leaf. The flowers and foliage appear about the commencement of the hot weather. They can be seen in their full beauty in April and May. Roxburgh and Brandis condemn the flowers as of a dirty grey or greenish-yellow colour, but although they may be unattractive from a distance on account of their diminutive size or entire absence of petals, and positively repulsive on account of the offensive odour as they open, their bright yellow anthers and the crimson streaks on the calyx, variegated with the soft down inside and outside the calyx, as clearly seen even under an ordinary magnifying glass, are by no means unattractive; at any rate they are not suggestive of any dirtiness in appearance. The offensive smell resembles that of *Sterculia foetida* or of *Sterculia guttata.* Male flowers are usually to be seen on the upper part of the spikes; they are sessile. The hermaphrodite flowers are chiefly confined to the lower part of the spike. They have short pedicels. Bracts linear; very early caducous; of a brownish colour; not seen at the base of the topmost flowers of the spikes. Sometimes two or three flowers arise from the same point or base on the spike.

CALYX. — Deciduous; free part of the calyx cup-shpaed, cleft half way into five triangular segments, woolly inside with long whitish-crimson hair; the hair outside is short.

ÆSTIVATION. — Valvate, Calyx-tube of the hermaphrodite flowers avove the ovary, with a campanulate mouth; segments of calyx pointed slightly backwards, i.e., towards the dorsal surface.

COROLLA. — Absent.

STAMENS. — 10.

FILAMENTS — 5 short, 5 long, arranged alternately, inserted below the calyx-segments; the larger ones twice the length of the calyx. In the hermaphrodite flowers there is an epigynous brownish disc between the stamens and pistil, densely hairy. The upper part of the filaments curves over the top part of the anther as it thins out into the connective.

ANTHERS. — Bright yellow, bold, slightly reniform with convexity on its ventral surface; consisting of two distinct loculi separated below, uniting at the connective; bursting longitudinally. The loculi are joined to the connective at their topmost part on their dorsal surface.

PISTIL —

STYLE. — Slender, filiform, tapering at the apex to half the size it has near the ovary, projecting slightly beyond the filaments.

STIGMA. — Simple, a mere depression at the apex, turning brown about the time the anthers burst and throw out their ample, impalpable offensive smelling pollen.

OVARY. — Inferior, coherent with the tube of the calyx, 1-celled, always tomentose.

OVULE. — Pendulous from the apex of the cavity.

FRUIT. — DeCandolle, Roxburgh, Wight and Arnott, Dymock, Lyon, & c., all agree in calling the fruit a "Drupe." Clarke in Hooker's Flora of British India discards this name. A *drupe*, according to the signification assigned to the term by modern Botanists, is morphologically a different fruit; and accustomed as I am to the arrangement and terminology of Bentley as regards Fruits, I think that the fruit of *Terminalia bellerica* being inferior, cannot strictly be called a drupe, which is a superior mono-carpellary fruit, that is to say, formed of a single flower, the ovary of which is made up of a single carpel. Clarke calls it *drupaceous*, and that is a more accurate term for it. The fruit is fleshy, indehiscent, $\frac{1}{2}$-$\frac{3}{4}$ inch in diameter, ovoid, grey or tan-coloured, velvety when young, turning into reddish or deep buff-colour as it matures, with five more or less distinct furrows $\frac{3}{4}$ to 1 inch long. The skin becomes coriaceous as the fruit grows older; epicarp and mesocarp shrivels as the fruit dries up after it is detached from the parent tree. The fruit ripens during the cold season. It is so persistent that the fruit which ripened last cold weather can even now be seen in the jungles in the middle of May on the trees just as they are throwing out the fresh foliage and flower-stalks of the current season.

NUT. — Thick and hard, rough, irregularly coarse grained, varying from $\frac{1}{2}$-$\frac{3}{4}$ inch in length; $\frac{1}{4}$-$\frac{1}{3}$ inch at its broadest part, which is about the middle.

SEED. — One, which constitutes the kernel in the nut. The nut yields a bland oil and is sweet to taste.

ALBUMEN. — None, says Clarke, as also Roxburgh.

COTYLEDONS. — Convolute; creamy white colour.

EMBRYO. — Inverse, spiral.

REMARKS.

By far the most accurate description I have seen of this plant is that given by Brandis (p. 222, Forest Flora of North-West and Central India). I have therefore tried to follow him and Wight and Arnott as much as possible in the description given above, and the remarks that are to follow.

There is one think striking about the venation of the leaves of the tree. Half-way between the midrib and the margin of the leaf, the main lateral nerves on either side of the midrib suddenly swell out into longish irregular nodules, assume a deeper crimson tint and develop more abundant hair. This is more marked on the older green leaves than on the earliest tender crimson ones. So far as I am aware this peculiarity has not been specially noted by any observer before.

The observation of Roxburgh, that there are two opposite glands on the upper side of the apex of the petiole and sometimes near the base, is not borne out by the experience of Wight and Arnott and Brandis. I have myself failed to find the glands Roxburgh speaks of, although I have examined several specimens in the plains and hills of Thana, and at Matheran. Clarke, in Hooker's Flora of British India, in referring to the supposed occurrence of glands on the leaves of *Terminalia bellerica* says that they are often on the petiole, *or near the base of the midrib beneath.* (The italics are mine. - K.R.K.) When clarke speaks of glands on midrib, and on the under surface of the leaf, just near the apex of the petiole, he was evidently thinking of or had before him the leaf of *Ain (Terminalia tomentosa)* instead of the leaf of *Beheda (Terminalia bellerica)*. I have a specimen of each of these plants before me as I am writing this paper.

DeCandolle says that the petioles are glabrous. Roxburgh also says the leaves are glabrous. It is not so. The whole organ is essentially tomentose; so is the flower; so is the fruit. An ordinary magnifying glass is enough to show the fine, white or buff coloured hairs or down on the petiole, midrib, nerves and their minuter ramifications; on the upper and under-surfaces of the leaf, several, weeks after the foliage appears. As the leaf gets older and assumes a brighter green, the down on the leaf shrivels. It may then disappear and perhaps the oldest and most mature leaves may appear glabrous. That is the only way I can account for the statements of DeCandolle and Roxburgh. The down on the petioles is distinctly of a crimson or coppery tint.

Roxburgh further observes that "the Natives do not use any part of the fruit of *Terminalia bellerica* in medicine." This remark appears to be based on scantly information. As a matter of fact, *Behedâ* is one of the commonest articles used not only by learned Hindu Vaidyas and native practitioneers of all sorts, but also as an ordinary domestic remedy. It is well known to the natives that *Triphalâ* is a penacea for all kinds of ailments.

Triphalâ is a combination of three fruits, *viz.* —

Hirdâ — *Terminalia chebula.*

Behedâ — *Terminalia bellerica,*

Awalâ - *Phyllanthus emblica,*

vide p. 44, Ark-Prakash, vol. 3, Nighanta Ratnakar.

These are pounded together, and used in the shape of a powder, decoction, or cold infusion. They are valuable for the large quantity of tannin they contain. They are valuable as astringents when given in combination. The mixture, however, acts as a laxative in children. It is generally known to give tone to the intestinal canal. It must be remembered that the part of the fruits used medicinally is the rind, not the pulp, as is wrongly said, for there is no pulp strictly so-called on any of these nuts covering the hard kernel. The kernel of the nut is never used medicinally. It is eaten for its sweet taste, sometimes in large quantities, as many as twenty-five being swallowed at a time without harm. I have heard from several reliable sources that the kernel is harmless. From other sources equally reliable I hear that if the kernel is eaten in large quantities it acts as an intoxicant. But of this more hereafter. Brandis says that the fruit of *Behedâ* is a favourite food of monkeys, deer, sheep, goats and cattle. The taste of the rind of the fruit is astringent. The taste of the kernel is like that of its congener *Badam (Terminalia catappa).* There is no evidence of the kernel of *T. catappa* ever having produced intoxication, eat what quantity of it you like. It is the tannin in the rind of the fruit that gives *Behedâ* its commercial value in common with its congener the *Hirdâ*, as an important ingredient, largely used in and out of India for dyeing cloth and leather and for tanning. *Beheda* is also known in the Konkan by the name of *Yelyâ.*

An insipid gum of brown colour exudes from wounds in the bark of *T. bellerica*, in vermicular pieces, about the thickness of the finger. "It is hardly at all soluble," says Dr. Dymock, "in water, in which it swells up and forms a bulky gelatinous mass." Roxburgh's statement , that the gum is soluble in water, is nearer the truth. I have verified it myself and have just at this minute a specimen of the gum in water, and I find it is as easily soluble as the ordinary gum arabic. Dr. Dymock was not able to verify from personal observation the statement of Drury that the gum burns like a candle. From my own experience I can say that it does burn and continues to burn with a pale reddish flame, and with a spurt and a flash now and then, swelling the whole mass out and leaving a charred residue. I have tried the experiment just this minute on a fresh bit of gum obtained from a bark wounded only yesterday. Perhaps Dr. Dymock's specimen of gum was old and dry, and had got rid of its inflammable material in the process of drying, as is but natural to suppose.

Wight and Arnott observe in their Prodromus that "the tree known as *Terminalia moluccana* of Roxburgh and probably of Blume (but not of Lamarck) is allied to *T. bellerica*. It has precisely the same kind of fruit and male flowers, but the petioles are short, giving quite a different aspect to the plant."

The flowers of *T. bellerica* are of offensive odour as I have already stated above. I have alluded to an incident at Thana regarding this offensive odour in a Lecture on Indian Flowers I delivered last year before the Sassoon Mechanics Institute of Bombay. Extracts from this lecture have already appeared in the last number of the Society's Journal wherein the incident is mentioned. I need not therefore repeat it here.

THE POISONOUS PROPERTIES.

The kernel of the seed of *Beheda* is narcotic in its action. Dr. Lyon refers in his Medical Jurisprudence (p. 222) to two sets of cases of accidental poisoning by it. One set of cases he quotes from Dr. Norman Chevers, Medical Jurisprudence as recorded by Mr. C.E. Raddock, Sub-Assistant Surgeon of the Malwa Bheel Corps, wherein three boys are mentioned as having suffered from poisonous symptoms after eating some of the dry nuts. They all recovered however. The symptoms in two of the boys were drowsiness, headache, sickness at stomach and free vomiting of a thick white frothy mucus. With regard to the third boy, who was weakly and seven years old, the symptoms were more severe. Briefly stated, they were as follows — Total insensibility; heart's action frequent and weak; legs cold, eyes rather glistering, and pupils normal but fixed; jaw firmly closed. This boy had eaten the largest quantity, it is said between 20 to 30 kernels. He is said to have played all day and night before going to bed. He showed no symptoms of poisoning till the following morning, when he was found insensible. Notice in these cases the long time that it took for the poisonous symptoms to show themselves. In the first two boys they did not appear till about eight hours after, and in the third boy not even in twelve hours.

The other set of cases referred to by Dr. Lyon are from the Bombay Chemical Analyser's Report (1878-79, p. 14), in which a woman and two children, one of whom was a weakly girl of eight or nine years, were poisoned. Of these, the girl died and the two others recovered. The symptoms were nausea and vomiting followed by narcotism. Dr. Lyon observes that it is not known to what these narcotic properties are due. In a collection of old Sanskrit observations on Indian drugs reçently edited by Mr. R.S. Lale under the name of *Guna Dosha Prakash* (1892, p. 135) the kernel of *Beheda* is referred to as intoxicating. O'Shaughnessy seems to have heard of this property.

From all accounts I have hitherto received, it appears probable that the eating of the kernel affects differet individuals differently. In some it produces intoxication; in others it is harmless even in large quantities. This reminds one of the effect of a variety of *Arecha catechu* or *Supâri* nut, known as *Mâjri Supâri* has on different individuals. *Mâjri* means "intoxicating". True to its name in some, *Mâjri Supâri* unmistakably produces giddiness, nausea, hiecup; whereas in others it is perfectly harmless.

In the printed Proceedings of the Grant Medical College Society for the year 1880 (*vide* Part No. 8, p. 71) there is an abstract of an interesting paper recording a case of poisoning by the kernel of *Behedâ* seed, which occurred in the practice of Dr. Tribhuvandas M.Shâh, L.M., then Assistant Surgeon at Ahmedabad, now Chief Medical Officer of Junagardh State. Dr. Tribhuvandas is one of the most accomplished Medical Graduates of the Bombay University. He is as careful a clinical observer as he is a skilful and facile operator. I cannot, therefore, do better than give in full, with his kind permission, the abstract of his notes as publised by the Grant College Medical Society — "A child, about five years of age, was brought one morning into the Hospital in an insensible condition, the parents stating that the child had eaten some of the kernels of the nut the previous evening. The parents were dyers and used the nuts for dyeing clothes black; the pericarp was used by them while the kernels were thrown away. The Natives are well aware of the effects produced by the eating of the kernels. About half a pound of the nuts will be eaten before they produce narcotic effect in an adult; the effect is not rapid but takes several hours before it is exhibited. The child had apparently gone to bed all right and in the morning was found quite insensible. When seen she was totally unconscious, the pupils were of natural size, the cornea was insensible to touch, breathing was stertorous, body of natural temperature, pulse somewhat frequent, jaws partially locked." Emetics, stimulants and other remedies were administered. The child, however, died of convulsions within half an hour from the time they set in. A *Post mortem* examination was made two hours after death. Herein lies the value of this paper. The stomach was found somewhat congested and contained some greenish black grumous fluid; the intestine and all the other organs were found natural.

The question that strikes me is — Is it that the symptoms of intoxication or narcotism may be due to small quantities of Hydrocyanic acid developed in the kernel of some of the fruits of *Beheda.* I leave the question to be solved by Practical Pharmacologists and Chemical Analysers.

Reprinted from
Journal Bombay Natural History Society
Vol. 8 pp. 331-334. 1893

PART VII.

STRYCHNOS NUX-VOMICA. — (*Linn.*)
(Natural order — LOGANIACEAE).
MARATHI — कुचला; काजरा KUCHALA, KAJARA

A tree over 40 feet in height, with a straight thick trunk.

ROOT. — Thick and with a yellowish epidermis; very bitter.

STEM. — Often 12 feet in circumference.

BRANCHES. — Dense, irregular, covered with a smooth ash-coloured bark. Young shoots shining and deep green, often tinged with red; bark nodose, bitter glabrous.

PETIOLE. — $\frac{1}{5}$ - $\frac{1}{2}$ inch long; deeply grooved.

STIPULES. — None, says Roxburgh in his "Coromandel Plants". Between the opposite leaves there is a raised line, which is perhaps a rudimentary interpetiolar stipule.

LEAVES. — Glabrous on both sides, shining, opposite, entire, coriaceous, often decussate in an oblique manner, arising from stout nodes; ovate or rotundate, sometimes elliptically oblong; 3-5 nerved; shortly acuminate or almost apiculate. 1-$\frac{1}{2}$ inches X 6 inches; usually 3-5 inches long. Base obtuse, somewhat unequal.

PEDUNCLE. — $\frac{1}{2}$-2 inches.

FLOWERS. — Many, small, greenish-white, appearing with young leaves on short slender pedicles; collected on small terminal pubescent corymbose cymes 1-2 inches in di.ameter, at the end of the branchlets or on short axillary shoots; pentamerous; bisexual.

CALYX. — 5 parted; persistent; $\frac{1}{2}$ or $\frac{1}{5}$ the size of thc corolla.

COROLLA. — Valvaate, hypogynous; regular, tubular or funnel-shaped, with a 5-lobed reflexed short limb. Tube $\frac{1}{4}$ to $\frac{1}{3}$ incvh long, glabrous at the throat, lobes valvate, about $\frac{1}{6}$ inch long; glabrous; a few conical hairs lower down the tube.

STAMENS. — 5, epipetalous in the throat of the corolla tube, alternating with the corolla segments.

FILAMENTS. — Scarcely any, or exceedingly short; inserted over the bottom of the division of the corolla.

ANTHERS. — Oblong, glabrous, half within the tube, half out.

STYLE. — Of the length of the corolla tube; glabrous; filiform.

STIGMA. — Small or short; undivided, capitate; sometimes indistinctly 2-lobed.

OVARY. — Free; 2-celled.

PLACENTAS. — Fleshy; adnate to both sides of the dissepiment.

FRUIT. — A berry, globose , smooth, indehiscent, with a fragile shell-like rind, orange-sized and orange-coloured when mature; filled with a soft white jelly-like pulp, which is intensely bitter.

SEEDS. — immersed in the pulp, 2-5 in number; $\frac{1}{2}$ inch in diameter, circular, discoid, shining, light grey, silky; not reniform as Brandis says, but having one surface convex, and the opposite correspondingly concave, with a small foveola in the centre of each side.

ALBUMEN. — white, horny or cartilaginous as Gaertener calls it.

EMBRYO. — Very small, compared with the size of the seed; Straight, ecentric; milk-white.

COTYLEDONS. — Cordate, acuminate, tri-nerved, very thin.

RADICLE. — Clavate, very small, placed near the hilum.

REMARKS

The wood is very hard and close-grained, white or grey, with numerous medullary rays. One cubic foot weighs 52 pounds. It is used for many purposes, such as ploughs, cart-wheels, cots and fancy cabinet work. The tree appears to be a native of Ceylon. My description is mainly drawn from the specimens obtained from the two handsome trees growing at Bassein in the Salsette Island in a garden near the ruins of the old Portuguese Fort.

Every part of the plant is exceedingly bitter, particularly the root. The pulp of the fruit, says Roxburgh, "seems perfectly innocent, as it is eaten greedily by many sorts of birds." Colonel Drury quotes this observation in his "Useful Plants of India." The root has the reputation of curing intermittent fevers. Rheede says that when boiled and drunk, it is purgative. The bark is used as an antidote for snake-bite. Brandis says that the pulp in the fruit is orange-coloured. It is not so; it is white. It is difficult to understand how such a careful observer as Brandis says so. It is evidently a misprint or slip of the pen.

The seeds contain 0.28 to 0.50 per cent. of an alkaloid called *Strychnia*, mixed with another alkaloid *Brucia*, closely related to it. *Igasuric acid*, similar to *malic acid*, is associated with these alkaloids. It is these alkaloids which render the plant poisonous.

The late Professor Sir Robert Christison says that the bark might be advantageously substituted for the seed in the preparation of strychnia.

The tree flowers in the cold season. Kurz in his "Forest Flora of British Burma." (vol. II. pp. 166-167), says it flowers in April and May. It may be so in Burma. The trees in Bassein flower in January. The fruit is ready in the early part of the cold season. Kurz says that the tree sheds leaves in the hot season. It is not known to do so in Salsette.

Brandis says the seeds are flat. If it be so, it is quite exceptional. The general form of the seed is correctly described by Gaertener when he calls it convexo-concave.

Roxburgh observes in his "Coromandel Plants" that the shell covering the fruit is somewhat hard. It is not so when mature and dry. It has the appearance of being so when the fruit is but half developed and the pulp has not yet become jelly-like, but is dense and comparatively drier. When, however, the fruit matures and the pulp is well formed and becomes almost isolated from the shell, the thinness is apparent. It is still more so when the fruit becomes dry; the seed and the pulp then lie loose in the cavity, and the shell easily cracks with a resinous fracture when pressed between the fingers.

Hooker adds a very important note in the description of this plant (*Vide* "Flora of British India," Vol. IV. p. 90), which is as follows — "Bentham also reduces to the present species *Strychnos ligustrina* (Blume, Rumphius I, 68, t. 25), which does not differ by any tangible character, but has smaller somewhat different-looking leaves."

POISONOUS PROPERTIES.

Nux-Vomica is so well-known for its poisonous properties that it is hardly necessary to do more in these pages than state them briefly.

Strychnia, the chief active principle of this plant, is one of the most powerful poisons acting on the nervous and muscular systems. It causes tetanus — that is to say, tonic contractions of all voluntary muscles. These contractions are generally sudden and last from a few seconds to many minutes. They follow each other in rapid succession. In severe forms there is hardly any intermission. The whole body in such cases becomes "rigid, immoveable, and hard as a board" (Schmiedeberg).

The convulsions excited by this alkaloid originate in the spinal cord probably by acting directly upon the motor-cells. The reflex irritability of the spinal cord, of the medulla oblongata, and of the brain is excessively increased. This cuases tetanus. When the brain and medulla oblongata are in this state, the spasms get excited by the slightest often imperceptible, stimuli, which may meet the eye, the ear, and particularly the organs of touch, so that they apparently come on without a cause (Schmiedeberg).

Strychnia has been found in blood. It has a marked effect on circulation. The blood pressure rises; there is arterial tension during the appearance of the convulsions; the frequency of the pulse becomes simultaneously slowed. This Mayer believes to be due to vaso-motor spasm from increased irritability of the origins of the vascular nerves and the cardiac inhibitory fibres of the vagus.

It must be remembered that the mind is perfectly clear in strchnia-poisoning. Strychnia is a cumulative poison. It also diminishes the process of oxidation in blood — that is to say, the amount of oxygen absorbed and of carbonic acid given out by blood are diminished (Harley).

Brucia is another alkaloid found in Nux-Vomica, but in smaller quantity than strychnia. It possesses properties similar to strychnia, but as a poison brucia is less active than strychnia.

Reprinted from
Journal Bombay Natural History Society
Vol. 8(4) : 453-461. 1893

PART VIII.

CERBERA THEVETIA — (*Willd.*)

(Natural Order — APOCYNACEAE.)

MARATHI — पिवली कण्हेर Pivli kanher

Synonym, THEVETIA NERIIFOLIA (JUSS). — It is known in England as the Exile tree. In this country it is known as the Yellow Oleander.

A handsome, ornamental, ever-green, spreading shrub, 12-15 ft. high. It may also be classed as a tree. Every part of it contains abundantly a thick acrid milky juice. The pith of the young branches is particularly rich in this juice.

STEM. — Rough, rugose, branching irregularly. Wood soft, whitish, and bitter.

BRANCHES. — Rounded, often rising in clusters; sometimes springing from the ground separately.

BARK. — 1-2 lines in thickness. The fresh bark of young wood is smooth, bright green, and covered by a thin grey rough epidermis. The epidermis can be readily scraped off; abundant milky juice escapes from the exposed green bark.

The wood becomes more compact as it grows older, and is then brownish, occasinally streaked black here and there. The pith becomes absorbed as the wood gets older, and leaves an irregular central canal, often large enough to admit of a crow-quill. The bark at this time has a brown suberous coating, much cracked here and there.

STIPULES. — Absent; there are several glands in the axilla of the leaf instead.

LEAVES. — Alternate, linear to narrow-linear ; 2-5 inches long, 1/4x inch at the broadest part, glabrous, coriaceous, glossy; usually veinless; in some, the primary veins which arise from the mid-rib are faintly prominent; acuminate at both ends, decurrent in a very short petiole, or almost sessile; margin revolute, entire.

FLOWERS. — Large, bright yellow, 2-3 inches long, streaked on the dorsum, with lanceolate twisted greenish stripes from the tube to the distal end of the corolla, terminating in a point.

INFLORESCENCE. — Terminal or lateral cymes. Peduncles much shorter than the leaves, rounded, glabrous, about a line in thickness.

ÆSTIVATONS. — Contorto- imbricate, the twist being to the right.

CALYX. — 5-parted; sepals distinct and persistent; lanceolate or ovate-lanceolate; glabrous, deep green externally, yellowish or light green internally; acuminate; $\frac{1}{3}$ the length of the corolla- tube, or about 3 lines long.

COROLLA. — Deciduous, falling-off the day after the opening of the flower; glabrous externally throughout, and internally as far as the throat of the tube, below which it is thickly covered with fine white hairs. The tubular portion is darker yellow inside, greener outside.

The form of the corolla is said to be hypocrateriform by De Candolle, as also by Mr. A. Smith in his article on this plant in "Lindley and Moore's Treasury of Botany." But the tube in this flower is neither long nor slender, nor is the limb, that is, the terminal lobes of the corolla, flat. The form of the flower is more campanulate than anything else. The corolla may be divided into three parts — (1) a tube about $\frac{1}{2}-\frac{3}{4}$ inch long, widest at its insertion on the calyx; about 2 lines in diameter, narrowest at its mid-part, and contracted at the throat; (2) beyond the throat is the bell-like central portion, $3/\frac{4}{1}$ inch in length; (3) from the distal part of this bell-like mid-portion five lobes twist in a spiral manner to the right, $3\text{-}\frac{4}{1}$ inch in length. The tube is all but closed in above the anthers by ovate acute- bilobate "*scales*" at the throat. These "*scales*" are covered with soft white shining hairs. They leave just a small opening enough to show the pointed apex of the stigma.

STAMENS. — 5, alternating with the lobes of the corolla.

Filaments. — Hardly perceptible.

Anthers. - 2-celled, included within the tube by the "*scales*" mentioned above; inserted on the throat of the tube; erect, introrse, subsessile, lanceolate; enclosing the stigma at its base by their inferior pointed extremities, almost adhering to the stigma.

Pollen. — Granular.

PISTIL —

OVARY — Superior, free, single; 2-celled. The entire ovary is surrounded by a bright yellow ring-like five notched disk, the notches alternating with the lobes of the calyx.

STYLE. — Filiform, included; each cell of the ovary sends up a distinct thread which unites with its fellow about a line away from the point of its exit from the cell.

STIGMA. — Conically capitate; faintly 2-lobed; the apex of the cone finely pointing at the distal end.

OVULES. — 2, in each ovary; amphitropous; round, compressed from base to apex; equi-distant from the ovary.

FRUIT. — Globose or ovately-globose, or half-orbicular; truncated at the apex; smooth glossy; light green when immature, yellow externally when mature; slightly fleshy; flesh white; longest diameter $1\frac{1}{2}$-2 inches horizontally; dilated at mid-portion transversely, with a hard brown stone or putamen which contains the seed. The putamen is of a peculiar triangular shape, resembling the "cocked- hat." It is divided by a deep groove into two cells along the edge, corresponding to the base of the triangle. Each cell is generally described as having a distinct seed within it, but very often one of the cells seems to remain undeveloped. In such cases the seed in such a cell is abortive.

SEEDS. — Two, often only one, slightly winged near the situation of the embryo; pale yellow; the tegmen of the seed is friable.

ALBUMEN. — Absent.

EMBRYO. — Erect, small, fleshy.

RADICLE. — Short, eccentric, cylindrical.

COTYLEDONS. — Rotundate; convex externally, plane internally; fleshy; varying in size from a two-anna silver-piece to a four- -anna silver-piece. The substance of the cotyledons is intensely bitter.

REMARKS.

The shrub is a native of South America and the West Indies. It was introduced into England in 1735. It is not known when it was introduced into this country and whether it came here direct from its native soil or from England. It is, however, perfectly naturalized in India in the different parts of the country. It has been figured and described by Royle in his "Illustrations of the Himalayan Plants" published in 1839. The flower is considered by some writers to have a sickly odour. It has the faint odour of *Nerium oleander*, which is by no means disagreeable. But that is a mere matter of taste or opinion; and, as such, knows no definite law. The wood also, as has been already observed, is credited by some writers of note, such as Lindley, with a repulsive odour ("Veget, Kingdom," p. 559, 1847). It does not possess any such odour, so far as I can judge from the examination of several specimens, fresh or dried, growing in Bombay, Thana, and Bandra, where the plant is much appreciated as a garden beauty of hardy growth.

Kurz, describing the plant as it grows in Burma, gives its height as from fifteen to twenty feet. It seldom grows as high as that on this side of India. The length of the clear stem is given by him as between six to ten feet, but very often here there appears to be no clear stem, as fresh branches are given off, even when the stem is old the branches keep sprouting out even down to the root. The flowering time in England, as given by Loudon, is from January to July ("Encyclopædia of Plants", p. 148, 1829). Kurz gives the flowering time in Burma as between November and March, where the fruit is ready in the cold season. The plant seems to flower and mature fruit on this side of India at all times of the year. Here it is quite possible to see a full-formed fruit, while fresh flowers are being thrown out in succession.

The latest example in proof of this may be given from the specimens I examined in January last in the Victoria Gardens, Bombay, through the kindness of Mr. C.D. Mahaluxmeewalla, the well-informed and energetic superintendent of the Gardens. In one of these specimens on a single branch, there was a mature fruit on one branchlet, on another a half-developed fruit, on the third a full-blown flower. In the Bandra specimens which I examined through the good offices of Mr. E.M. Pratt, I.C.S., in July and August, 1892, and in January last, there were flowers and fruits in all stages of development on one and the same tree.

The plant is said to require rich mould when grown as a stove-plant in cold countries. In India it grows in red soil, as well as in sandy or clayey soil.

With regard to the question of stipules, Alph. De Candolle has the following important observation in his monograph on the *Apocynaceae* in "Annales des Sc. Nat." (3rd Series Vol. I., Bot., pp. 235 263) — "In *C. thevetia* there are axillary fasciculated glands, that is to say, accumulated in large numbers, between the stem and the petiole. Considering their situation, number, and smallness, one cannot in any way take these glands for stipules." (J. Mittra.)

Dr. Norman Chevers says in his celebrated work on Indian Medical Jurisprudence that the oil obtained by expression from the seed of *Cerbera thevetia* is amber-coloured and slightly viscid, and that it is recommended as a cathartic by natives; "but," he adds on the authority of Dr. Shortt, "it produces violent vomiting and hypercatharsis." Col. Drury observes in his "Useful Plants of India" that the oil is of a bright yellow colour, but that its uses and properties are yeat undetermined. Dr. Dymock and his colleagues in their "Pharmacographia Indica" (Vol. III, pp. 406-410), as our latest authorities on the subject, say that the oil when pure is as inert as olive oil. The oil, according to the researches of De Vrij, has an agreeable mild taste like that of fresh almonds. De Vrij says that the oil is limpid and almost colourless. The quantity that can be obtained by expression is 35-41 per cent., and 57 per cent, with benzol.

POISONOUS PROPERTIES

The plants of the order *Apocynaceae* abound in milky juice containing various acrid principles. The acrid principles isolated from

Cerbera thevetia, according to the most recent researches, are known as Thevetin and Theveresin. These active principles render the plant highly dangerous. It must not, however, be supposed that all the plants of the order *Apocynaceae* are of such a nature. Owing to the absence of one or other of the acrid principles, the milky juice of a plant belonging to this order may be perfectly innocuous. Royle mentions, on the authority of Lindley, a plant of this order which is perfectly harmless. It is the milk-tree or *Hya-hya* of Demerara, called by Mr. Arnott *Tabernaemontana utilis*. Royle mentions other harmless plants, such as the *Melodinus monogynus*, which is a native of Northern and Eastern India, and the orange-like fruits of which have a sweet edible pulp; the *Carissa edulis* of Nubia; and the Cream fruit of Sierra Leone; all these yield edible fruit. The *Willoughbia edulis* growing in India and the *Carissa carandas* growing wild all over the Thana District, are instances of plants bearing perfectly harmless fruit. There are, on the other hand, other plants of the same order, such as *Cerbera Manghas*. and *Cerbera Tanghin*, which are highly dangerous. The seeds of *Cerbera Manghas* are narcotic. The seeds of *Cerbera Tanghin*. (the Tanghin tree of Madagascar) are said to be so powerfully posionous that a single seed may be sufficient to destroy *twenty* persons (Royle).

The bark of the plant, administered internally, has been, since the days of Lindley, considered to be a powerful febrifuge, 2 grains of it only being considered to be equal to a dose of cinchona. It is used in countries where it grows wild for poisoning fish.

I must not omit to mention here the observations of Mr. A. Smith in the "Treasury of Botany", which are as follows — "Its fruit is regarded as noxious, though, according to Dr. Seeman, a gentleman ate four of them, when a boy, without injury." Possibly this was due to idiosyncrasy, by virtue of which the boy's constitution withstood the effects of the poison, or that he ate the quantity of the fruit on a loaded stomach, which possibly diluted the poisonous elem- ent to such an extent as to render it inert. Again, we are told that it was the fruit that was eaten. We are not aware to what extent the fruit, i.e., the pulpy part, contains the poisonous element, nor do we know that the pulp is ever used for extracting the poisonous principles. Perhaps the poisonous principles exist in much less proportions in the fruit-pulp than in the seeds, which latter is the part of the plant generally used for the extraction of the active principles. We have clear instances in recorded cases of how deleteriously the active principles of the plant may sometimes act on certain individuals. It is the milk of the tree that is considered *highly venomous*. Its bitter bark and leaves are cathartic.

Dr. Norman Chevers, on the authority of Dr. Waring of unrivalled therapeutic repute as regards Indian plants, observes in his Medical Jurisprudence that the kernels of the plant are exceedingly bitter. I know it by personal experience, for I have tasted them and found them so to my disgust. "They produce," says Dr. Norman Chevers, "a slight feeling of numbneess, very slight indeed, in the tongue;" but I did not notice any heat whatsoever which Dr. Waring says is experienced.

From a case published by Dr. J. Balfour ("Madras Journ. of Lit. and Sc." 1857, Vol. III., N.S., p. 140), it appears that the swallowing of one kernel of the fruit produced acro-narcotic symptoms, similar to those of aconite. In a case quoted by Dr. Chevers, as mentioned to him by Babu Kannya Lall Dey, a case which occurred in the practice of a physician whom the latter knew, it appears that on the eating of a single seed, the following symptoms occurred — In a short time after swallowing, vomiting commenced, but there was no purging; within half an hour the boy, who had swallowed it, was covered with cold clammy sweat; the countenance was pale, and the eyes sunken deeply, and within two hours he died convulsed.

Let us now consider the active principles of the plant under notice. They are Thevetin and Theveresin. Dr. Dymock and his colleagues in their "Pharmacographia Indica," elaborately quote the experiments of T. Husemann on the lower animals, such as frogs, dogs and rabbits. Our concern is more with regard to man, and it will, therefore, be more useful to summarize the researches of a leading Pharmacologist like Schmiedeberg, so far as they could be determined on the mammalian genus *Homo*, with regard to the action of the poisonous group to which Thevetin belongs.

Schmiedeberg classes. Thevetin with the Digitaline Group. The active principles of the various plants, classed under this group, act upon all sorts of animals in so uniform a manner that each of the substances seems, as regars its individual action, a true copy of any one of the others. The most important of this group is the one which gives it its distinguishing name in pharmacology, namely, Digitalin, which, together with Digitalein and Digitoxin, form the active principles of Foxglove. Next to these comes the active principle of Squills, named Scillain. Next come Helleborin, the active principle of Hellebore, and Oleandrin, otherwise called Nerein, to which the Common Oleander owes its poisonous properties. Then there is Apocyanein. Finally, belonging to this groups are Adonis, the active principle of Adonis, Vernalis and Convallamarin, the active principle of *Convallaria Majalias* (Lilly of the Valley).

Scillain, Helleborin, Adonis, and Oleandrin or Nerein are not crystalline and are almost insoluble in water. Digitalin, Apocyanein and Convallamarin are glucosides which form frothy solution with water like Saponin, a glucoside contained in the bark of Quillaia of Chili (*Quillaia saponaria*). Thevetin, Antiarin and Strophanthin, belonging to the Digitaline Group, are considered by Schmiedeberg as rarities.

The action of these active principles included in the digitaline group consists almost exclusively in the fact that they alter in a peculiar way the elasticity of the cardiac muscle without at first modifying the contractility of it. There are three stages distinctly, from their preliminary action to their most potent final effect : (I) — *In the first stage*, there is a rise of blood- pressure and consequent slowing of the pulse, larger quantities of blood being thrown into the aorta and thence into the arteries. This effect can be kept up by small doses of the substances of this group. (II) — *In the second stage*, there is continuous high blood-pressure with irregularity both in the action of the heart and in the frequency of the pulse. (III)— *In the third stage*, there is sudden arrest of the action of the heart from the paralysis of that organ as indicated by a rapid fall of the blood-pressure.

It is as such that Thevetin is supposed to act, especially producing the phenomena characteristic of the second and third stages. The subcutaneous injections of this substance, Schmiedeberg observes "cause phlegmonous inflammations more or less readily." Closely connected with this, is its action on the stomach and intestines which causes gastric disturbance, diarrhoea and other symptoms well known to have been observed during the use of digitaline. The action of Thevetin appears, from the recorded cases of poisoning, to be more rapid than in cases where digitalin is administered medicinally.

These views of Schmiedeberg are corroborated by Carna. His experiments show that Thevetin produces death by asphyxia and cardiac paralysis. Applied to the skin, says he, it produces irritation and a sensation of burning. It produces also convulsions of cerebral and paralysis of spinal origin. It increases intestinal peristalsis, lowers the temperature, and, locally applied, it increases salivation, and contracts the pupils. Warden has confirmed the statement as to the production of convulsions. (Dymock).

As against what Schmiedeberg has said — namely, that Thevetin is non-crystalline — De Vrij says that Thevetin is obtained at the rate of four percent from the cake left after the oil is expressed from the seed. It consists of a "beautiful white crystallized glucoside." This question is well worthy of the study and further research on the part of the Indian pharmacologist. De Vrij was able to find Thevetin in the bark also of *Cerbera thevetia*. Pure Thevetin, says Dr. Warden of Calcutta, is white and *Crystalline* ("Pharma, Journal," Vol. XIII, 3rd Series, p. 43). Warden has discovered a second poisonous principle in the seeds of *Cerbera thevetia*, to which he gives no name, but it is one which appears to him to possess greater activity than Thevetin. Pure Thevetin, he says is only faintly bitter, whereas the newly discovered poisonous principle is persistently bitter and *non-crystalline*. The latter was obtained by precipitating the mother-liquour left after the crystallization of Thevetin by aqueous tannic acid and decomposing the precipitate by lime. In his work on Medical Jurisprudence, Dr. Lyon observes that in India cases of poisoning by *Cerbera thevetia* in the human subject are seldom met with. But Dr. K.L. Dey says that he has lately found it employed in Bengal for homicidal purposes without raising any suspicion. Two cases of homicidal poisoning came under his observation in the course of a month. Nine cases of cattle poisoning by the seeds of this plant came under the observation of Dr. Lyon in 1886. Dr. Dumortier records a fatal case in a child three years old after swallowing one seed.

I may observe here that Dr. K.L. Dey considers the oil obtained from the seeds of *Cerbera thevetia* to be a cardiac poison. He says that "the acrid oil contained in the kernels is a powerful acro-narcotic poison allied to Nux Vomica, and its effects as a poison are very rapid" (Pharmaceutical Journal, Vol. XII, 3rd Series, p. 397).

Reprinted from
Journal Bombay Natural History Society
Vol. 9(1): pp. 42-60. 1894

PART IX.

AMORPHOPHALLUS CAMPANULATUS — (*Blume*)

(Natural Order — AROIDEAE)

MARATHI — सुरण SURAN

An annual, tuberous, herbaceous, seedless plant, flowering before leafing. It has no nectaries.

SYNONYM. — ARUM CHAMPANULATUM (Roxb.)

TUBER. — Larger in size than the adult human head; sometimes three times as large in the cultivated variety; smaller than the adult human head usually in the wild variety; spheroid, with concentric tuberculate rings on the upper surface, where on the central part there is a shallow concavity or depression, often a slight circular prominence round the scape, corresponding to the attachment of the petiole of the previous year. After the leaf has completed the maturation of the new tuber, it separates, petiole and all, from this depression or prominence which is from 1-1/2 to 2 inches in diameter and has at its central part, a mammilla marking the spot whence the flower-stalk of the future year shoot forth in the succeeding hot weather.

The substance of the tuber is hard, solid, fleshy; externally pale blackish; internally white, pale pink, or deep brown; on section discharging drops of acrid, clear, shining, gummy mucus, which turns brown on standing; the tuber is studded with numerous rootlets. The general appearance of the tuber presents a marked convexity on the lower surface, and a slight concavity on the upper surface; sometimes the upper surface is almost flat.

N.B. — Properly speaking, the underground stem we have named as Tuber here on the authority of Hooker is a *Corm.*

ROOTLETS – Long, filiform, whitish, denser near the scape, beyond which they are thicker, often of the size of a goose-quill, arising here there in large numbers from small bubils or tubercles from all parts of the main tuber. Such is the habit of the tuber that if by any process of arrest of development, such as rotting, the entire formation of the tuber is interfered with, we find a number of small tubers of varying size and form instead of a whole spheroidal tuber. These small tubers are generally united; sometimes detached. Whether united or detached, however, each one of these small tubers may produce a separate plant if placed under favourable circumstances, such as soft soil and proper manure. From such detached tubers, generally there is no flower-stalk thrown out before the leaf makes its appearance in the succeeding rainy season. Thus, from such a large disorganized tuber of the previous year, I have been able to obtain in my garden several small but healthy plants, each springing from a partially separated or entirely disjointed bulbil. It seems to me that when tubers are small, flower-stalks very seldom spring forth.

THE PETIOLE. — I call it *the petiole*, for there is, as a general rule, *only one* given out by a properly formed tuber. There may be two or even more, but they are the result of improper development of the tuber as noted above, *i.e.*, where there are several bulils of large size. The rule is, one well-formed tuber, one scape, one petiole and leaf. Another petiole or many more petioles may shoot forth from large or small bulbils. But this is a distinct deformity in the plant. The petiole of a fairly well- developed plant is radical; rough or warty externally; slightly sheathing near its attachment to the tuber, between the margins of which is a space for the mammilla of the future year's flower- stalk, varying in length (or rather in height) from $1\frac{1}{2}$ to $2\frac{1}{2}$ feet; sometimes in the cultivated variety, as grown in my garden, as much as 4 feet. In the wild variety it is hardly ever more than 2 feet. In thickness it is usually $2\frac{1}{2}$ to 4 inches in diameter, cylindrical, softly fleshy, succulent. Colour varying from light to deep green, mottled with irregularly shaped whitish, pale bluish spots; the petiole, dividing into three parts, radiately diverges in a horizontal manner. The divisions are dichotomous, pinnatisected, with a deep channel on their ventral aspect, roundish on their dorsal aspect, slightly scabrous. There are membranaceous wings along the sides of the decurrent bases of the segments of the leaf. These segments are lanceolate or ovate-lanceolate, the margin entire, parallel veined, pale green and entirely glabrous.

SCAPE. — Solitary, 2 to 4 inches long, cylindrical, greenish, mottled with white spots, very thickly verrucose, invested below with two or three imbricated scales or bracts which are linear- lanceolate, tough, fleshy, rose-coloured, mottled with green or purple spots.

SPATHE. — Is technically called "*marescent*," *i.e.*, not actually falling off before the spadix is perfected, but withering long before that time. In size very large; ovate; 1 to $1\frac{1}{2}$ ft. long; very broad; erect; acuminate; below, of fleshy substance, infundibular, convolute; above, membranaceous, campanulate, patulous, with undulate curled margins. The convoluted part in its greatesh circumference is about a foot and a half; externally, speckled with pale yellowish-greenish patches and bright green dots; internally, at the base it is purple with very thick fleshy warts, which are thickest and deepest coloured near the scape, and which become paler coloured and less dense as they approach the mid-part of the infundibular portion. The mid-part is conspicuously greenish-yellowish, without any warts. The colour of the undulate marginal portion is purplish or dark purple; that on the under or external surface is dull; that on the upper or internal surface is bright.

SPADIX. — Projecting distinctly beyond the spathe; erect, thick, club-shaped, almost half-way from below cylindrical and pistil-bearing; thence upward it is pear-shaped and thick, bearing anthers; above this part lies the apex (appendage or club) expanding in a globosely conoid irregularly formed mass when young, which becomes fungating and sinuously lobed as it matures. The texture internally is spongy, fibrous, lacunose; externally corrugated, brownish dark purple, resembling soft leather with minute warts or projections alternating in regular order with shallow depressions between. As the conoidal apex matures into the more corrugated mass of sinuous small lobules they emit an intolerably offensive odour of putrid flesh, inviting hordes of blue-bottles and other large flies which cover the whole mass with their eggs; and the subsequent maggots, which thickly beset if for the next four or five days, render the flower stalk as disgusting to the eye and nose as carrion. I have already referred to this fact in my lecture on Indian Flowers delivered before the Sassoon Mechanics Institute, Bombay, extracts from which have already appeared in one of the previous numbers of this Journal (*Vol. vii, Part 4, pp.* 527, 528 — 1892).

FLOWERS - Unisexual; the males on the middle third of spadix, immediately below the appendage; the females on the lower thirdor basal part of the spadix. Perianth absent. Male and female flowers contiguous *i.e.*, having no neuters between.

MALE FLOWERS. —

Stamens. — Numerous, dense.

Anthers. - 2-celled, sessile, close-packed, consequently compressedly cylindrical; rounded at the top; straw-coloured; glabrous.

Connective. — Longitudinal, fleshy separating the two loculi. Each loculus perforated at its apex for the discharge of the pollen by a central pore, which is at first semi-lunar, then gradually dilating to assume a roundish form.

Pollen. — Globose; lemon-yellow, or almost orange-coloured.

FEMALE FLOWERS —

Pistils. -Numerous; not so close-packed as the stamens, but somewhat loosely and spirally arranged; the topmost scarcely separated from the anthers by any clear space.

Stigma. -Trilobed, often bilobed; all on the same level; much thickened and expanded. There is a deep depression between each lobe; slightly papillose. Cream-coloured or yellowish.

Style.- Many times longer than the ovary. Crimson or purplish. $\frac{1}{4}$ to $\frac{1}{3}$ inch long; deciduous.

Ovary. — 2-, rarely 3-celled, with only one ovule; purple or deep crimson; broad and dwarf; globose or somewhat cubical; fleshy glabrous.

Ovules. — Attached to the inner angle of the carpel at its base; ascending, obovoid, anatropous.

FRUIT. — Not yet known to have ever developed.

REMARKS

The above description is mainly based on Blume's as embodied in Rumphia (vol. I, p. 139), and that of Roxburgh and of Sir Joseph Hooker, in the Coromandel Plants (vol. III, p. 70), and Flora of British India (vol. VI, p. 513), respectively. I have added, however, what I know personally regarding this very interesting species of the Aroid family. The propagation of this plant takes place by means of its tubers. It is hardly necessary, therefore, for the ovule to develop into seed. The main tuber or each small bulbil or tuber of the previous year serves as the parent of the future plant. The flower-stalk seldom, if ever, appears in young bulbils or even in tubers half as big as the adult human head. I have come across several tubers of even larger sizes where the flower stalk has not appeared, but in due season only the petiole has sprouted out. Blume observes that the scape is almost twice as thick as the petiole. In my experience it is quite the reverse, but this may be merely a local peculiarity.

With regard to the size of the spathe sprouting from very large-sized tubers. I have this to observe. In tubers weighing from 30 to 40 lbs., the spathe assumes an enormous size, and its terminal convolute portion and crimpled margin measure several feet in circumference. I have now before me a large tuber which, when fresh removed from the ground, weighed fully 40 lbs. The convolute portion of the spathe thrown out by this tuber measured 8 ft.; in circumference; the height of the spathe was near 2-1/2 ft.; the diameter of the middle campanulate portion was 1 foot; the broadest part of the appendage was nearly of the same diameter; the diameter of the staminal part was nine inches, whereas that of the pistil-bearing part was six inches. This shows to what an enormous size the tuber and the flower-stalk may develop under cultivation. In the uncultivated or wild variety, which goes under the name of Ran-Suran or Jangli-Suran, the plant is not known to develop to such a size. The wild variety in every part of it is altogether a much smaller plant.

It must be noted here that the wild or uncultivated variety of *Amorphophallus campanulatus* is entirely different from the tuber of *Amorphophallus sylvaticus** (Kunth), which is also locally named Jungli-Suran and described by Dr. Dymock under that name in the *Pharmaceutical Journal* (p. 172, vol. vii, 3rd Series). Dr. Dymock was well aware of this fact at the time he first described this plant, for he has since repeated his remarks in his later works, - viz., "The Vegetable Materia Medica of Western India" and "The Pharmacographia Indica" — that "it is probable the two plants are distinct," for he adds that "it is probable the roots of the wild *Amorphophallus campanulatus* form a part of the commercial article known as *Madanmast*, which is the tuberous root of *Amorphophallus sylvaticus* peeled, cut into slices and strung upon a string."

* SYNONYMS – (I) *Synantherias sylcatica* (Schott.) - Hooker's Flora Br. Ind., part XIX, page 517; (II) *Arum silvaticum* (Warden and Pedler's article on the nature of the toxic principle of the *Aroideæ* – Journal As. Soc., Bengal, vol. LVII, part 2, No. 1)

I now proceed to state what the old Sanskrit writers have said regarding the varieties of *Amorphophallus campanulatus*.

Madanpal, in his Nighant (Pushpa-Varga, Shlokas 83 to 86), says that there are two varieties of Suran: the tuber of the one is round and well shaped, that of the other is irregular in form. The former is the better of the two; the latter is known as Ran- Suran and has also the epithet of Vajrakand. This epithet, as applied here, means that the tuber of this variety is as irritating as Vajra, which is a synonym of Kush or Darbha, a kind of grass described by Linnaeus as *Poa cynosuroides* (*Eragrostis cynosuroides*, Rom. et Sch. according to modern nomenclature). Now, it is well known that the blades of this grass, which is very commonly used in the religious ceremonies of the Hindus, are covered with long stiff hairs, especially at their margins; these hairs produce an unpleasant, prickling sensation on the skin of persons whose tactile sense is acute. The pricking sensation produced on the mucous membrane of the mouth, throat and tongue of those who eat the tuber of the wild variety of Suran is very much like that of Darbh grass on a sensitive human skin. The epithet of Vajrakand as given to ran- Suran would therefore seem to be appropriate. I have so far dilated on this point as I find Dr. Dymock gives "Thunderbolt" as the English equivalent of Vajrakand (Pharmacographia Indica, vol. III, p. 547). Vajra does mean in Sanskrit thunderbolt; but if Dr. Dymock's English rendering is to be accepted, one cannot find any connection between the term thunderbolt and the shape or properties of the tuber of Ran-Suran, unless it is meant to convey that its poisonous quality is as suddenly destructive as the thunderbolt.

Narhar Pandit in his Râjh-Nighant, another standard Sanskrit work, also mentions two varieties : (1) Suran; (2) Sit or Swet Suran, which is the white variety.

Dhanvantari Nighant, another old Sanskrit work, mentions only one variety of Suran. It is probably the cultivated form, as the author does not make any reference to the wild variety.

Bhâv Misra, another standard writer, in his celebrated work Bhav Prakash (p. 150, Bombay Jagadishwar Press Edition, 1888), speaks of only one variety. From its description it appears to be the cultivated form, as no mention is made of the wild tuber.

Let me now proceed to notice a few points worthy of special remarks in the writings of European botanists who have studied the plant under notice. Wight, in his Icones Plantarum Ind. Orient. (vol. III), gives two plates, numbered 782 and 785. The former illustrates a gigantic flower-stalk with its enclosing spathe, and the latter depicts the tuber, leaf and flower -stalk. He says that he depicts the gigantic flower-stalk, as he finds no reference to a flower of such size in Roxburgh's work. Botanists have to be thankful to Wight for this illustration, as it serves to persons like myself of a younger generation to emphasize what he observed in his days, by adding my own humble testimony that his picture illustrates but a gigantic, well-developed inflorescence of one and the same species as given in his plate No. 785. This I have been able to observe for a number of years in the flower-stalks thrown out by large tubers weighing 30 to 40 lbs. The larger the tuber, the more gigantic the flower. By some unfortunate mistake, however, (*evidently that of the printer's devil*), Wight's descriptions of his plate 782 and 785 are transposed in the letter-press given at the beginning of the third volume of his Icones, which at first sight is confusing to one who has not had the opportunity of watching the growth of gigantic flower-stalks from very large-sized tubers. The result is that even a learned and veteran botanist like Sir Joseph Hooker, in quoting Wight in his "Flora of British India" (*vide page* 513, part XIX), is obliged to put a query against Wight's plate 782*. From my own observation, however, I am able to say — and I Say it with profound respect for Sir Joseph Hooker — that Wight's plate No. 782 represents an exceptionally large flower-stalk, obtained under cultivation of a large tuber of the very same plant depicted by Wight in his plate No. 785 of the same work. If hereby I am able to satisfy Sir Joseph Hooker, I shall have done some service to Dr. Wight, however humble, in having tried to remove a doubt as regards the accuracy of his observations — a doubt which, I admit, must necessarily arise in the minds of careful investigators of the stamp of Sir Joseph Hooker wholly and solely from the unfortunate transposition of Dr. Wight's description of his plates.

* I address the remarks to Sir Joseph Hooker, although I know from his note at p. 490, part XIX of his Flora of British India that the Descriptions of the Indian Species of the Aroid Family are drawn up for his work by Mr. N.E. Brown.

It is not clear what Dr. Wight means, in the letter-press alluded to above, by saying that the flowers are "sessile with respect to the surface of the ground." for be it noted that there is a distinct scape whether the tuber is underground or aboveground at the time the spathe sprouts out. The spathe gives cover to the spadix on which, from below upward, are the three distinct parts *viz.*, I, the pistil-bearing and II, staminiferous portions, and above them, III, the lobulated deep purple appendage. Blume supports me in this statement, for in his Rumphia (vol, I. p. 141, 1835), he describes a distinct scape which shoots forth from the tuber *aboveground* (the italics are mine — K.R.K.) before the rains, when the tuber is entirely destitute of its large solitary leaf or smaller solitary leaves from its superimposed bubils.

Paxton, in his Botanical Dictionary (revised by Hereman, 1868), says that *Amorphophallus campanulatus* is a native of Ceylon. I may add that it is a native of Western India as well, especially of the Konkan.

With regard to the action of light on the flower-stalk of this plant, I may be permitted to observe as follows — when the spadix sprouts out in a room where the direct rays of the sun do not reach, the purple colour is very poorly developed, it is very faint. A marked difference is noticed in the depth of the colour directly such a pale growing flower-stalk with its tuber, is bodily removed from a dark room to one where it can enjoy the direct rays of an Indian sun. The pale purple is at once changed into a deep bright purple which might shame the Tyrian purple of ancient note.

Some botanists say that the plant is perennial. It is a mistake to suppose that either the plant or the tuber is perennial. It appears to be strictly an annual plant. The following is its life-history so far as I have observed it during the last ten years, if not more. Given a tuber of the previous year, which matures here between October and November; in March or April, or even as late as the end of May, the flower-stalk appears; this takes about a month to develp and throw out the pollen; ovulation, or fruit-forming, seldom, if ever takes place in this part of India; the spathe, the spadix and the male and female flowers then gradually wither away; the peduncle shrivels just before the rains, and as it is bodily falling off from the tuber, the leaf-bud of the future year's plant sprouts out not very far from the base of the shrunken flower-stalk. All this may happen aboveground, should the tubers be preserved in a store-room; or if they be underground, the flower-stalk makes its appearance aboveground and goes through the same process of development. At the beginning of the rains the tubers have to be put into the ground. The softer the ground the better is the chance of obtaining a big uniformly well-developed tuber at the end of the rains. With the rains the plant grows quickly, waving its tripartite solitary leaf three or four feet above ground. The chief nourishment of the leaf is the starchy matter of the old tuber; as this is absorbed by the petiole and carried upward into the farthest ramifications of the leaflet, the old tuber vanishes. The leaf when mature begins about the middle of the rains to replace its old used-up tuber by the gradual formation of a new one. This new tuber is ready about the end of the rains, or the beginning of the cold weather. That this new tuber is a totally different individual from the old one is proved by the fact that where, by some error in the proper performance of the function of the leaf, *viz.*, that of using-up the old starch and reforming a fresh quantity, the leafy expansion does not use-up the whole tuber of the previous year, the new tuber that is formed can be seen as quite a separate individual superimposed on the old half-used-up tuber in the same way as we find corms placed one upon the other — the new avove the old-in the natural order *Iridaceae.* As soon as the growth of the tuber is completed, or rather as soon as the power of starch-formation in the leafy expansion is exhausted and the process is at an end, the leaf and petiole begin to turn yellow shortly after the rains, shrivel and dry-up, and ultimately get bodily detached from the new tuber, leaving a scar in the form of a depression or prominence, from the central part of which the flower-stalk of the future plant springs forth. Such is the life-history of this interesting plant.

POISONOUS PROPERTIES

The cultivated variety of Suran has for a long time been used by the native of this country for culinary purposes, although it is not free from acridity even after boiling. The acrid principles are to a very large extent got rid of by soaking in water the cut pieces, of the tuber intended for culinary use. Even then the tuber does not cease to be irritating to the mucous membrane of the mouth, throat, and tongue. Some may ask why a tuber which is one of the recognized vegetables of the Indian kitchen and one of the most palatable of them when properly salted, spiced, and cooked (boiled or fried in ghee or butter), should be classed under the poisonous plants of Bombay. My reply is, the wild and the cultivated tubers are so difficult to distinguish, and the wild variety is so vilely irritating to the digestive tract, and even the cultivated variety is so frequently productive of irritation in the mount, throat, and tongue, that it is safer to look upon the plant with suspicion even in its highly cultivated forms. For, although cultivation does to a very large extent reduce the acrid principles to a minimum, they can never be said to be altogether absent from the tubers, and cases have been met with where a sensitive throat has considerably suffered from irritation for some time after partaking of a dish of the cut pieces of even a cultivated tuber of Suran. Some throats may not suffer at all; others may suffer more or less. It is safe to be forewarned.

The poisonous properties of Suran were not unknown to the old Sanskrit writers of repute. Both the varieties mentioned in Madan-pal's Nighant are said to be productive of *itching* (Sanskrit — *Kandu*). The same property is attributed to the two varieties mentioned in the Raj Nighant of Narhar Pandit. No reference, however, is made to any irritant or poisonous property in the single variety mentioned in the Dhanvantari Nighant, probably because, as I have already observed, the writer was describing a cultivated variety in which the itching or irritant propety is invariably less marked. The single variety mentioned in Bhav Prakash, however, is distinctly noted as being possessed of irritant property. So far, as regards what is known from ancient indigenous writers regarding the obnoxious nature of the plant.

Let us now see what the researches of European & Native experts have done in Europe and India to settle the question of the poisonous properties of the Aroid family in general, and the Suran species in particular. I must refer the reader at this stage to the elaborate and valuable investigations of Dr. Warden and Mr. Pedler of Calcutta, as embodied in their article already referred to, as a contribution to the Journal of the Asiatic Society of Bengal. They say that from the brief resume given by them of the Arums found in India, a belief in the toxic properties of certain species would appear to be pretty generally entertained. Dr. Murrell's remarks as published by him in the British Medical Journal (May 7th, 1881), and referred to by Warden and Pedler regarding *Arum maculatum* (Cuckoo- pint) were not the first to point out the poisonous nature of some species of the Arum family. Sowerby and Johnson describe the plant, in their work entitled the British Poisonous Plants published so far back as 1861, as possessed of a powerful acrid principle. They note that in two or three cases leaves eaten by children produced disastrous effects. The fact that the root also of *Arum maculatum* possesses acrid properties when bitten was not unknown to them. Miss Annie Pratt had also referred to them in her illustrated pamphlet on the British Poisonous Plants long before the researches of Murrell or Warden and Pedler. *Arum maculatum* appears to be the only British representative of the Aroid family. In India we have several species developing tubers possessed of similar, more or less acrid properties. Nine of these have been stated by Warden and Pedler to be distinctly poisonous.

Sowerby and Johnson observe that "the poisonous qualities of *Arum maculatum* are wholly dissipated by heat and the plant then becomes highly nutritious." They further add that the roots of various foreigh species are employed as food after being macerated in water, baked and reduced to powder. I cannot say that in every case the acridity of even the cultivated varieties of Suran *entirely* disappears on boiling. I know of no instance where in this country Suran is ever reduced to powder or baked before use. It is invariably used after boiling in ghee or butter, or mixed with molasses.

Guy and Ferrier, in their Forensic Medicine, make an important observation which is worth noting here. They say that the juice of *Arum maculatum*, when applied to the tongue, causes darting pain as if it were pierced with sharp needles. The same sensation is noticed by some after eating Suran even when well boiled. The sharp needle-like ends of the crystals (technically called *raphides*) found in the cellular tissue of Suran, and more fully described hereafter, whatever their chemical nature, would seem to account for their irritant property.

Prof. Christison observes in his work on Poisons (p. 602) that he knows from personal observation that when the roots are distilledwith water neither the distilled water nor the residue possesses acridity. The fact that the distilled water possesses no acridity would point to the crystalline raphides being insoluble in water; the residue possessing no acridity would point to the fact that the acicular raphides under the action of heat so change their form as not to be very sharply pointed and capable of entering the mucous surface of the mouth, tongue, and throat, and in consequence become inert. This remark of Professor Christison's is very important, inasmuch as we find that, if pieces of the tuber of Suran are eaten after boiling them well, they may at times be rendered free from severe poisonous effects, even although there may be a slight irritation of the throat, mouth, or tongue. Some persons may not feel this at all. Professor Christison quotes an observation made by Reinsch, which although it refers to *Arum maculatum*, may well be applied to Suran, viz., that, though the powder of the root may not be acrid to taste, it yet produces severe burning of the throat not long after swallowing it. Woodman and Tidy say that, in addition to the symptoms just mentioned, *Arum maculatum* is known to produce dilated pupils, insensibility, and coma. They refer to a case recorded by Dr. Stubé (*Lancet*, 13th April, 1872), in which a man took a leaf to get rid of tapeworms. He suffered "from immediate pain and pricking sensation in the mouth downwards; the tongue became swollen, and there was salivation and vomiting." Profuse salivation has been noted by Dr. Dymock as occurring in a congener — *Synantherias Sylvatica*, Schott (*vide* Pharm. Ind., vol. III, p. 547). The case mentioned by Dr. Frayer (Br. Med. Journ., 22nd January, 1861), is even more characteristic, as there was a spasmodic action of all the muscles of the body, bloody froth at the mouth, dilatation of the pupils, very feeble action of the heart, and rigid closure of the jaw. These severe symptoms have never been known in the case of Suran, although it is quite possible they might develop, the nature of the irritant being identical. In another case, recorded by Dr. Frayer, of poisoning by *Arum maculatum*, there was burning pain in the mouth and lips, torpor in three hours, followed by complete prostration in six hours, with delirium, asphyxia and death in nine hours (Warden and Pedler). In a case recorded by Dr. Alliott (Br. Med. Journ., 23rd April, 1881), there was vomiting and severe purging. This is not know to have occurred in any case of Suran-poisoning.

Beck, in his, Medical Jurisprudence (p.940, ed.1836), refers to a member of the Aroid family mentioned by Hooker inhis Exotic Flora, the poisonous properties of which are very similar to those of *Amorphophallus campanulatus.* It is the *Caladium sanguineum,* Vent, of the West Indies, where it is called *Dumb-cane* from the fact that its virulent juice, when applied to the tongue, causes a swelling which deprives the sufferer of the power of speech. Woodman and Tidy think that this species contains a large amount of prussic acid, as two drams of the juice have proved fatal in a few hours. Beck mentions another similar Aroid, the *Caladium arborescens,* also a native of the West Indies, which, on the authority of Merat, he says, is so caustic that occasionally the lips of Negroes are wetted with it as a punishment for slight misdemeanors. The poisonous action of both these *Caladiums* is narcotico-acrid. The action of several other plants of the Aroid family mentioned by Beck, such as *Arum maculatum,* Ln., *Arum dracunculus* L., *Arum dracontium,* L., and *Arum triphyllum,* L., is that of an irritant poison. They are all acrid, and said to be dangerous. Beck quotes from Orfila's Toxicology (vol. II, p. 83) a case related by Bulliard, which is as follows — "Three children ate of the leaves of *Arum maculatum.* They were seized with horrible convulsions, and with two of them all assistance was unavailing, as they could not be made to swallow anything. They died — one at the expiration of twelve days, and another at the expiration of sixteen. The third was saved with difficulty. Its tongue was greatly swelled, and hence deglutition was painful and difficult."

Arum fornicatum (Syn. for *Alocasia fornicata*), the *Bish Kachoo* of Bengal, is an equally poisonous member of the Aroid family. Warden and Pedler refer to a case noted by the Chemical Examiner of Bengal in which the tubers of this species produced symptoms of irritant poisoning. In the experiments they made with some tubers locally known as *Bish Kachoo,* which were sent to them by the Civil Surgeon of Dibrugarh, and which Dr. King thought were most probably those of a species of *Alocasia* or *Colocasia,* they arrived at conclusions of which the following is a summary. They think that the active principle is non-volatile, as there was a complete absence of any irritative action on the olfactory organs or conjuctivorae, though there was considerable irritation of the hands while handling the tubers during their experiments. A minute fragment of the tuber applied to the tongue caused in a very short time acute lancinating pain, which continued for a considerable period. In their experiments with a viscid alcoholic extract obtained from exhausting the powder of air-dried tubers with hot 60 O.P. alcohol, and driving off the alcohol by the heat of a water-bath, they found that the extract was without the least action on the tongue. An extract obtained with cold alcohol in which the alcohol had been driven off by spontaneous evaporation, similarly produced no action on the tongue. A glycerin and an ethereal extract likewise yielded negative results. The effect of distilling fresh tubers with water was that the distillate had no acrid taste; it contained traces of prussic acid. The tubers left in the retort after distillation were, however, still physiologically active, indicating that the active principle was not dissipated by mere boiling with water. Warden and Pedler here observe that native using Arum for culinary purposes frequently add an acid vegetable or fruit, such as tamarind, to get rid of the acrid principle. This fact is well known in Western India; large quantities of tamarind are used in cooking the leaves of *Colocasia antiquorum* (the common Alu of Bombay) to keep down its acrid taste. In writing in a previous number of this Journal (vol. VII, p. 315), on the irritant action of *Pythonium Wallichianum (Shewla)*, another member of the Aroid family, I have already referred to the use of the fruit of *Bilimbi (Averrhoa bilimbi)* as an effectual and necessary adjunct in the culinary preparations of the flower stalks of *Shewla*.

Warden and Pedler observe, as already noted above, that they have no reason to doubt that the Arums as a class contain a toxic principle. They say that "while drying the tubers without artificial heat deprived them practically of all activity; exposing them to the temperature of boiling water for at least half an hour, at the most only very slightly diminished their activity." They note that in the American variety of *Arum maculatum* D.S. Jones has proved the presence of a volatile acrid principle besides starch, sugar, gum, albumen, resin, fat and extractives. They also say that Bird believes in the existence of a volatile alkaloid in it. I find that Taylor also says that the irritant properties of *Arum maculatum* depend on a volatile principle, as they are lost on desiccation or distillation with water (Poisons, p. 510, ed. 1848). Enz., in 1858, is said to have obtained *Saponin* from it. Now, it is well known from the researches of Schmiedeberg that the *Saponins* are amorphous or crystalline glucosides. They are found widely distributed in the vegetable kingdom, and are all, with but few exceptions, very soluble in water and render it frothy, Their action is related to that of *Emetine*. Injected into the blood, they cause paralysis of the central nervous system. "Small doses cause symptoms in man closely resembling the nauseant stage of emesis-that is, a raw, feeling in the throat, tendency to cough, and increased secretion of mucus" (Elements of Pharmacology, p. 68). According to Lauder Brunton *Saponin* produces anaesthesia, and even both sensory and motor paralysis, if applied locally (Pharmacology Therap, and Mat. Med., p. 836, ed. 1885). It may therefore, be presumed that it is questionable if *Saponin* is at all to be found in the species of the Aroid family we have been referring to. Paralysis has very seldom, if ever, been noted in cases of Arum-poisoning; whereas instead of there being anaesthesia, there is positive itching and hyperaesthesia.

But to return once more to the researches of Warden and Pedler. Their analysis of the ash of the tubers of Bish Kachoo indicated the presence of a large amount of potassium and magnesium; calcium was also present, but not sodium. The acids found were carbonic, phosphoric, and hydrochloric with traces of sulphuric. They also obtained marked quantities of potassic nitrate from the dried tubers. They add that the examination of the ash failed to afford them any clue to the physiological action of the fresh tubers. Their microscopical examination of a section of the tuber revealed numerous bundles of needle-shaped crystals. Similar crystals were also found in the leaves and stems. These crystals were insoluble in cold acetio acid, but easily soluble in cold dilute nitric or hydrochloric acid; caustic potash was without action. Crystalline bundles were visible in sections of boiled tubers, examined cold. Warden and Pedler come to the conclusion that these are crystals of oxalate of lime, and that the physiological symptoms caused by Arums are due purely to the mechanical irritation which they cause. "Bearing in mind the action of the re-agents on calcic oxalate," they say, "the reaon why mere boiling the tubers in water failed to deprive them of their activity is explained by the insolubility of oxalate of lime in water." Again, the action of dilute acetic acid even at a temperature of 100°C. in slightly lessening the activity of the tubers is due to the very slight solubility of oxalate of lime in that acid. And, lastly, the complete loss of all physiological action when the tubers were treated with dilute nitric or hydrochloric acid is evidently due to the ready solubility of calcic oxalate in these mineral acids. This was verified in every instance under the microscope. Warden and Pedler found that, on drying, the tubers practicaly lost the whole of their physiolgical activity. "Clearly," they say, "there could have been no loss of oxalate of lime on desiccation;" and as a matter of fact they found as many crystals under the microscope in the dried arum as in the fresh tubers. They admit however that this is an anomaly and explain it as follows — "In the fresh condition of the tubers, the bundles of crystals of oxalate of lime are cone-shaped more or less, the sharp points covering a wide area and forming the base, but in the drying of the tubers the needles appear to arrange themselves more or less parallel to one another, and the sharp points thus cover a small area. And thus, instead of each crystal acting as a separate source of Irritation and penetrating the tissues, the bundles act as a whole."

In the Pharmacographia Indica (vol. III, p. 547) needle-shaped crystals are mentioned by Dr. Dymock as forming the contents of the cells composing the parenchyma of the fruit and seed of *Synantheria sylvatica*, Schott.

The mechanical theory of irritation by the pointed ends of the crystals, say Warden and Pedler, has been independently adopted by Herr Stahl of Jena. They admit, however, that, in addition to local irritation, *viz.*, of the tongue, throat, and mouth, there are other and graver symptoms produced by the ingestion of Arum- tubers and leaves, such as convulsions, dilated pupils, insensibility and coma, and that the occurrence of these symptoms cannot be explained by the mechanical theory. They look upon these symptoms as the result of the action of the calcic oxalate crystals on the stomach. They say that the stomach at first suffers from mechanical irritation by the crystals; the secretion of the gastric juice thereby diminishes or ceases entirely; free hydrochloric acid of the gastric juice is, in consequence, not available for dissolving the calcic oxalate crystals; the crystals thus may pass on to the intestinal canal; once in the intestines, they may induce convulsions, dilated pupils, and coma by mechanical irritation in the same manner as intestinal worms.

Believing that the irritation caused by the tubers of Suran is due to the mechanical action of the needle-shaped crystals found in their cells in close-packed bundles, I have thus freely and fully reproduced, although in the form of a summary, the exhaustive researches of Warden and Pedler. The importance of thier results will be amply apparent from what I now proceed to record for the first time, as the result of a careful examination of the tubers of the wild and cultivated varieties of *Amorphophallus campanulatus* by another careful investigator hailing from this side of India. If the naming of the needle-shaped crystals is different in the following note, it will I hope, be considered, not that I reject the results of Warden and Pedler and accept those embodied in the observations which follow, but that I wish it to be clearly understood that, although the remarks of Warden and Pedler may apply to the Arums generally, the crystals found in Suran may not be calcic oxalates, but calcic sulphates and carbonates.

At my request, Dr. Nasarvanji Fakirji Surveyor, M.D., B.SC., M.A., M.R.C.P., a distinguished Graduate of the Bombay University, whose special training under renowned scientific workers in London, and whose marked ability and aptitude for microscopical and clinical investigation add a value of no ordinary kind to his remarks, has recently examined pieces of the wild and cultivated varieties of *Amorphophallus campanulatus*. The results of his observations are embodied in the following note which he has kindly communicated to me — "The Chief points of interest about their microscopical and chemical differences are as follows —

"The wild variety shows two forms of crystals, while the cultivated variety shows only one form. I shall first describe those crystals which are found in both the varieties and then those found in the wild variety only. Those found in the former I shall call the *white* crystals; while those found in the latter or wild variety only will be described as *brown* crystals. The white crystals are about 120* in length and 3 u in thickness. They are acicular and glistening and have a double contour. They are sharp-pointed at either extremity and always straight. They appear to be rigid; more numerous in the older parts, while they are almost absent from the youngest part (e.g., leaf-bud). They are insoluble in cold water; slightly soluble in boiling water. Acetic acid dissolves them with evolution of gas (CO_2 probably).

"On incinerating a thin film of the juice on a slide, these crystals were found to be fractured in numerous places , while many appeared to be either transversely striated or granular. This was due perhaps to the fact that the water of crystallization was driven out by the heat. These crystals were found to be scattered about, not collected in definite bundles. They were probably carbonate of calcium.

* μ = 1/1000 of a Millimetre

"The *brown* variety was only found, as has been already remarked, in the wild *Amorphophallus*. These crystals were very fine; about one-third the size of the first. They were also acicular, but did not show a double contour. They were found in the cells arranged in sheaves, and were distinctly brown when viewed in a mass. On adding a drop of water to the juice of the tuber, these cells swelled up and discharged the crystals. When examined singly, the crystals appeared to be very thin and somewhat curved. The brown tinge was still noticeable. The length of each crystal is 50 u, and thickness under $\frac{1}{4}$ μ. The crystals are sparingly soluble in warm water, freely in boiling water, so that a piece of the wild variety, on being reduced to pulp and boiled with distilled water, was almost deprived of the crystals, while the white crystals were still seen. By filtering the hot water, brown crystals were obtained from the filtrate by evaporation. Is it a fact that the wild variety can be eaten, if well boiled, especially with salt ?

"On incinerating the piece just as described above, I found that these crystals were not altered in the slightest manner, thus proving that they too were inorganic in nature and that they were not oxalates.

"Acetic acid dissolved them without evoluton of gas. These were probably sulphate of calcium crystals; however, I am not certain of that. I examined pieces of both the varieties for other salts aftert charring and incinerating them. This method showed insoluble and soluble carbonates, soluble sulphates and traces of chlorides.

"Of course, sulphate of calcium, being a poisonous salt, must have something to do with the properties of the wild variety; however, it is perhaps not the only cause; a glucoside or an alkaloid may be present too. As I have neither the means nor the time at my disposal, I am not able to settle that point.

"The important facts, however, are that the brown crystals are only present, as far as I know, in the wild variety, and that they are soluble in boiling water and weak acids without evolving any gas like the white crystals. These are very delicate, and appear in places to be slightly bent; while the white crystals are rigid, straight, and thicker."

I am very thankful to Dr. Nasarvanji for these remarks. They will serve as an important guide, incomplete though they are, in understanding the poisonous nature of the plant. With regard to his query whether "the wild variety can be eaten, if well boiled, especially with salt," I may say that the wild variety can be rendered partially (for I cannot say *entirely*) innocuous when well boiled with or without salt, but more so when boiled with acid fruits and vegetables. Generally the wild variety is not made use of for culinary, but only for medicinal purposes under the name of *Madan-Mast* as already observed. The wild variety is used as food in exceptionally hard times such as a famine. In the Dekkan famine of 1877-78 the leaf and tuber of the wild variety were used as food in the Khandeish District, probably not without boiling nor without the addition of salt. I say probably, because the wild tuber is very acrid, and under no circumstances can it be solely used without intense irritation of the mouth, throat, and tongue. In the cultivated plant too, there is sometimes a good deal of acridity, which, though not noticeable at the time of eating, is distinctly recognized shortly after eating. It even irritates the skin of the finger- tips as it is being cut by delicate female hands for culinary purposes in Hindu homes. Of this there is no doubt. Much of this acridity of the cultivated variety is got rid of by boiling. The slight persistant acridity would seem to be due to the presence of the stiffer and bolder while crystals which Dr. Nasarvanji pointedly observes are present even when the pulp of the wild variety is boiled with distilled water and the brown crystals are completely dissolved.

It is to be hoped that Dr. Nasarvanji may some day be able to complete his investigation of the tubers of Suran, or that some qualified chemical analyst may take the matter up and ascertain the exact nature of the needle-shaped crystals. That they act mechanically there is not the least doubt; guided by the researches of Warden and Pedler, this can be the only conclusion one can arrive at. In my own practice I have seen cases of distinct irritation of the throat, mouth, and tongue after the use of boiled Suran. In the practice of Dr. Sakharam Arjun and Dr. Vinayak Govind Gidh, swelling of the tongue and salivation were noticed in addition.

Reprinted from
Journal Bombay Natural History Society
Vol. 9(2): pp. 147-175. 1894

PART X.

MORINGA PTERYGOSPERMA — (GÆRT)
Natural order — MORINGEÆ.
MARATHI — शेगट, शेवगा SHEGAT, SHEVAGA

The tree is commonly known as the horse-radish tree. Some call it the drum-stick tree. Sir George Birdwood calls it the *smooth* horse radish tree.

ROOT. — Pungent; young parts tomentose or puberulous.

BARK. — Soft, gray, corky, whitish-green covered with a fine brownish epidermis; innermost part white; external surface reticulate, about one inch thick; deeply cracked; the bark of the young branches of some varieties of this tree is distinctly crimson; the ventral surface of some petioles is in such cases tinged with light crimson; the dorsal surface on the other hand is light green.

WOOD. — Soft, light, coarse-grained, spongy, perishable, white; rather light brown says Jaikisson Indraji. Young wood and branches very brittle and break with almost a resinous fracture.

From incisions, made on the trunk by the human hand or by insects which infest the bark, there exudes a pinkish, reddish gum in vermicular tears.

LEAVES. — Alternate, decompound, 1 to 2 ft. long, 2 to 3-odd-pinnate; usually 3-odd-pinnate near the extremities of branches.

PINNÆ. — 4 to 6 pair, opposite; the lower 3 to 4 pair, bipinnate.

PINNULÆ. — Opposite, 6 to 9 pair; the four lower pair generally with 3 to 7 leaflets; the rest consisting of single leaflets.

LEAFLETS. - 6 to 9 pair, opposite, $\frac{1}{2}$ to $\frac{3}{4}$ inch, quite entire, ovate or obovate, blunt; lateral nerves indistinct, glaucous- green above, pale beneath, slightly pubescent along the midrib beneath, caducous as well as the pinnules.

PETIOLE. — Slender, sheathing at the base.

PETIOLULE. — Slender, short.

Note that the petioles, petiolules, and leaves are all articulated at base.— (Baillon).

STIPULES. — Absent; glands instead, linear, hairy, stalked. These glands are sometimes also found at the origin of the petiolules and leaflets.

INFLORESCENCE. — Axillary; in numerous panicled racemes; spreading at the ends of branches. Baillon says that the inflorescence "consists of very numerous flowers in much ramified racemes of cymes."

BRACTS. — Linear, shorter than pedicels.

ÆSTIVATION. — Slightly imbricate.

FLOWERS. — Very numerous, rather large, white, yellowish-white, or reddish-white; irregular, hermaphrodite, pedicelled, 1 inch in diameter, honey-scented; pedicels 6 to 8 lines long. Baillon observes that "the receptacle of the flower is cup-shaped, linedby a glandular disk with a prominentx free border; on its oblique mouth are supported the perianth and androecium, while the gynoecium springs from the bottom." ("Natural History of Plants," Vol. III, 1874).

CALYX. — Cup-shaped, 5-cleft.

SEPALS. — Distinctly petaloid, unequal, linear-lanceolate, reflexed, deciduous; sometimes tinged pink and white, sometimes greenish white. Baillon observes that "the praefloration of the sepals is quincuncial." (Vide *Supplementary Plate* L.) The five alternating petals are usually of cochlear-imbricate praefloration. The anterior, internal in the bud, and differing in form slightly from the rest, remaining erect on anthesis, while they become reflexed on the receptacle like sepals.

COROLLA. — Petals free, 5 unequal, linear, narrowly spathylate; according to Hooker the upper petals are smaller. Lindley describes them as "ascending;" the lateral petals are ascending; the anterior are larger. Le Maout and Decaisne describe the petals as follows — "Five inserted on the calyx, linear, oblong, the two *posterior* (the italics are mine. — K.R.K.), rather the longest, ascending, imbricate in bud."

ANDRŒCIUM —

STAMENS. — Perigynous, unequal, inserted on the edge of the disk; declinate; in two whorls of five each; the outer whorl sterile, being reduced to mere slender antherless filaments; the inner whorl fertile, callous and villous at the base, superposed to or opposite the petals. Hooker and Brandis say that the sterile stamens are sometimes seven in number. Note that the sterile stamens are superposed to or opposite the sepals. Le Maout and Decaisne say that the stamens are 8 to 10, and line the base of the calyx. Baillon observes that "the androecium is most developed towards the anterior side of the flower, so that the fertile stamen superposed to the enveloped anterior petal is the longest of all; the posterior pair are the shortest." (*Vide Supplementary Plate* L.)

FILAMENTS. — Flattened and hairy at the base, free below, and above; slightly petaloid, connivent, united above the middle, unequal; the posterior ones longest; those opposite, the calyx segments shorter and imperfect.

ANTHERS. — Yellow, occasionally reddish; dorsifixed, consequently introrse; one-celled or simple as Lindley calls them, dehiscing longitudinally. Brandis describes them as oblong; Wight and Arnott describe them as peltate; Le Maout and Decaisne describe them as ovoid -oblong.

CONNECTIVE. — Thick, convex.

POLLEN. — Ovoid with three folds; in water it becomes spherical with three papillose bands (H. Mohl. *Ann. Sc. Nat. Ser.* 2, III, 343, -quoted by Baillon).

DISK. — Concave; lining inside of calyx-tube, with a short free margin. Baillon, as referred to already, describe the disk as glandular; the colour of the disk is green.

GYNŒCIUM —

OVARY — Superior, stiputate, hairy, lanceolate, one-celled, but composed of three carpels.

STYLE. — Slender, filiform, terminal, cylindrical, tubular (Baillon); obliquely recurved, but not obliquely inserted (Wight and Arnott); undilated at its stigmatiferous apex

STIGMA. — Perforated, simple (Lindley). Le Maout and Decaisne say it is truncate; Kurz says it is perforate and truncate.

OVULES. — Numerous, suspended or descending, anatropous; with their micropyle upwards and inwards; arranged in two rows, on three parietal placentas; raphe ventral and very prominent.

Baillon observes that "in the one-celled ovary there are three parietal placentas, whereof two are posterior." He adds a note, however, that we do find two or four carpels with the same number of placentas and valves to the fruit. I am quite familiar with this tree, having seen it during the last thirty years, if not more, growing wherever I have lived, and I can confidently say that I have not yet seen a two-carpellary or four-carpellary ovary in this plant either in Bombay or in Thana.

FRUIT. — A capsule, elongate 9 to 18 inches long, sometimes a little over two feet in length, 1/2 to 1 inch thick, rostrate, pendulous, torulose, 3 to 6-angled, with nine longitudinal ribs, 3-carpelled, 1-celled; loculicidally 3 -valved, valves bearing the seed along the middle of their inner surface. The capsule is succulent and soft within when unripe; corky and pitted when mature and about to dehisce; the outer coat of the capsule, which is chiefly formed of the ribs referred to above, consists of tough fibrous plates longitudinally, oftener torulosely, arranged and bound together by a green substantial parenchymatous substance.

SEEDS. — Trigonous, numerous in pits of the valves, "half-buried in their fungous substance *produced from the inner wall of the pericarp*" (Baillon). Hamilton * says that each capsule when full-grown contains about fifteen seeds. In size the seed is considerably larger than a pea. *Chalaza*, apical, corky (Le Maout and Decaisne).

TESTA. — Corky, winged at the angles, Baillon thus describes this winged nature of the testa from which the plant derives its characteristic name of *pterygosperma* — "In *M. pterygosperma* the superficial seed coat is hypertrophied into a vertical wing, which extends into the sinus between the two valves. These wings are imbricated with those of their neighbouring seeds, which may at maturing be arranged in a single vertical row along the axis of the capsule." "The number of wings," adds Baillon, "will vary of course with that of the valves, and also because the wing may remain exceptionally rudimentary on one or two angles of the seed, or not be formed at all, as is normally the case in *M. aptera*, an African, Arabian, and Syrian species."

* Pharmaceutical Journal, Vol. V, pp. 58 and 59, 1845-46.

ALBUMEN. — Absent.

EMBRYO. — Straight, "large, fleshy, oily" (Baillon). "Large, fleshy, occupying the whole interior of the seed" (Lubbock's "Seedlings")

COTYLEDONS. — Plano- convex, almond-shaped, fleshy.

RADICLE. — Very short, superior, turned to the hilum, included between the bases of the cotyledons and lying close to the micropyle at the hilar end of the seed.

PLUMULE. - Many-leaved. Sir John Lubbock observes in his work on Seedlings that "the plumule is well developed, and shows the primary leaves in a compound form" ("Seedling," Vol.I, p. 381).

REMARKS.

The following are the synonyms of the plant as given by Wight and Arnott ("Prodromus," p. 178):-

M. Oleifera, Lam. Encycl. Meth. 1, p. 398.

M. Zeylanica, Pers. Syn. 1, p. 460.

Guilandina Moringa, Linn, sp. p. 546; Roxb. in E.I.C. Mus., t.58.

Hyperanthera moringa, * Vahl. Symb. 1, p. 30.
Spr. Syst. 2, p. 327.
Roxb., Fl. Ind. 2, p. 368.

Anoma moringa†, Lour. - Rheede Malb. 6, t. 11.
Rumph : Amb. 1, t. 74.

Hooker gives the following in addition ("Flora Br. Ind." Vol. I) —

M. Polygona, D.C. Prod. II, 478.
Hyperanthera decandra, Willd.

It must be noted that *M. polygona*, D.C., is considered by Wight and Arnott (Prodromus) as perhaps a mere variety of the *M. pterygosperma* we have been describing here. They add, however, that they have never seen a specimen of the former. The character of *M. polygona* is described by them thus — "All the stamens bearing anthers; capsules many-angled; seeds 3- angled, the angles expanding into wings" D.C. (Prodr. 2, p. 478). The synonym of this also is given by them as *Anoma moringa*, Lour. (Burm. Zeyl., t. 75; Rumph. Amb. 1, t. 75). Note that *Anoma moringa*, Lour., as

* Also mentioned by Griffith in Notulæ IV 572, t. 609

† Willd. 2.536: Asiat. Res. II, 168. (Stamens half the number fertile, half sterile).

mentioned above is also the synonym they have given to *M. pterygosperma*; so that practically *M. polygona*, D.C., and *M. pterygosperma* must be considered the species of the same plant. The species with wingless seeds known as *Moringa aptera*, which is a native of Arabia and Africa, is not very distinct from *M. pterygosperma*. Sir George Birdwood observes that DeCandolle doubts the distinctness of these two species; " and it may be," say he, "that the Indian *Moringa*, although it has not the hard wood and oil seed of the Arabian, African, and West Indian Plants, is yet specifically identical with them."

The species described by Dalzell and Gibson as *M. concanensis*, Nimmo (in Graham's Catalogue, p. 43) would appear to be a mere local variety of *M. pterygosperma*. It is thus described — A tree something like *M. pterygosperma*, but distinguished by very much larger and rounder leaflets, and a much more powerful odour of horse-radish; flowers yellowish, with pinkish streaks; anthers 5, perfect but only one-celled; 5 abortive and much smaller; flowers in November in the Jungles near Penn (Kolaba Collectorate). Dalzell and Gibson have found it also on the hills in Lus, and have expressed an opinion that it very probably will be found in Arabia.

Nairne, our latest writer on the Flora of Western India, in referring to *Moringa concanensis*, observes that the plant is like the *M. pterygosperma*, but that its leaves and panicles are larger; its leaflets also larger oval, roundish; flowers sweet-scented; petals yellowish, red-streaked at the base. Nairne further remarks that the *M. Concanensis* is wild at various places in the Konkan, also in Sind and Rajputana. That it grows wild in Rajputana is evidently asserted on the authority of Brandis, who, as I have already noted in my former papers, is one of the most accurate observes and descriptive writers on the Indian Flora. It is just as well to quote his remarks *in extenso* — "*M. concanensis* grows wild on the dry hills of Rajputana (*Sainjna*) near Kishengurh and Bednore, also on the hills of Sindh (Mhua) and the Konkan. It has yellow flowers, bipinnate leaves, pinnae 4 to 6 pair simple pinnate, except sometimes the lowest, which are bipinnate, with broadly ovate leaflets one inch long, on petioles 1/3 inch long, with 4 to 6 pairs of rather prominent lateral nerves."

As a rule *Moringa pterygosperma* ismore largely found near human habitation-on the outskirts of existing or extinct villages, in cultivated fields and in gardens-than in the forest. But that it also grows wild cannot be doubted in the face of authoritative testimony. For instance, T.N. Mukharji, F.L.S., of Calcutta, an accomplished and well-informed Botanist, in his Hand-book of Indian Products (p. 98, 1883), remarks that *M. pterygosperma* is found wild in the sub-Himalayan regions.

Wherever it is cultivated, it is useful for its root, bark, gum, leaves, flowers and fruit. There is no part of this interesting plant but that it is used for culinary purposes. Every part of the plant is more or less pungent, the root notably so, and it may be added that every one of them is used to a more or less extent for medicinal purposes, and not without success. The root is mostly, I should say principally, used as a garnish at the Anglo-Indian table, as a substitute for the genuine European horse-radish, the *Cochlearia armoracia* (N.O. *Cruciferae*). Opinions however differ with regard to the flavour of the root of *M. pterygosperma* as an acceptable substitute for the genuine horse-radish root. Firminger says that as a substitute for horse radish it is much inferior in flavour. Surgeon-General Balfour* says that the roots of *M. pterygosperma* are *precisely* the flavour of horse-radish. Dr. Hamilton† says that the roots of *M. pterygosperma have all the flavour and properties of the horse-radish.* Dr. O'Shaughnessy‡ observes that the roots *have precisely the flavour of horse radish and seen identical in chemical composition (the italics are mine. - K.R.K.). Dr. J. Bidie of Madras regards M. pterygosperma* as a perfect substitute for *C. armoracia* — the true horse-radish. To this I may add my own humble testimony, having constantly tasted in England the true *C. armoracia* root as a garnish, that the root of *M. pterygosperma* is by no means inferior in pungency to the root of *C. armoracia.*

Passing next to the bark of *M. pterygosperma*, it may be observed that, apart from its noxious properties, to be dilated upon hereafter, it yields, says T.N. Mukherji, a coarse fibre for the manufacture of mats, paper, and rough cordage. From the very fragile nature of the bark I have very grave doubts as to whether it could afford useful

* Balfour's Encyclopædia of India, Vol. II, p. 986, 1885.

† Pharm. Journal Vo. V, 1845-46, pp. 58 and 59.

‡ The Bengal Dispensatory and Pharmacopœdia, Vol. I, p. 289, 1884.

cordage, notwithstanding the reticulate nature of its fibres; yet I can safely recommend it to the serious consideration of our indigenous paper-manufacturers as worthy of trial in the manufacture of coarse paper. I am all the more strengthened to make this remark, inasmuch as the plant is a very quick grower. Not only are seedlings said to be quick- growers, so quick as to bear flower and fruit within a year, but they are alos easily propagated. Dr. Hamilton says that the tree is of most rapid growth, coming into flower within a few months after the seed has been sown, and continuing to produce seeds and blossoms afterwards throughout the year. I have no personal knowledge of this plant being ever raised from seeds in Bombay or in the Thana District, which is the northern part of the Konkan; but I know this, that the plant takes so kindly to the soil, especially where there is plenty of water, as to give me the impression that it is easier to propagate it by cuttings than by sowing its seed. A branch, full two inches thick in diameter, if put into ground in the early part of the monsoon, which generally begins in Thana in the first week of June, by April following the cutting gives substantial fruit. I have Firminger's authority to say that even the seedling is of exceedingly rapid growth. He recommends the seed to be sown in June and July.

As regards the timber of this tree, Dr. Hamilton notes that it gives out a blue colour in spirit and water which, by transmitted light, appears of a golden yellow; the blue is destroyed by acids which leave the tincture or decoction of a bright yellow, but is restored by the addition of an alkali.

Next, as regards the height of the tree, it is variously given. Brandis says it is twenty feet. The trunk, he says, is straight; its girth from four to five feet, and there are a few large divergent branches. Kurz reckons the height as from thirty to forty feet, which is about the correct estimate I think. The height of the clear stem, according to Kurz, is from ten to twenty feet, and its girth from three to four feet. I may observe, however, that both in Bombay and Thana, the tree very seldom shows so much height in a clear stem. It is often irrgular in appearance; in some places markedly so. Dalzell and Gibson call it a small tree. Drury, again, is nearer the known height. He says that the tree is from thirty to thirty-five feet high.

There are differences of opinion regarding the nature of the stipules As a rule, theyare, as stated above, absent. Hooker, Kurz,Baillon and others say that, instead of the stipules, there are glands. Le Maout and Decaisne say the stipules are deciduous. Henfrey and Bentley say they are thin and deciduous. Lindley adds that they are coloured besides.

With regard to leaves, Jaikisson notes that they are acrid and have a burning taste. With regard to the shedding time of the leaves, Brandis observes that the old leaves are shed in December and January. The leaves are shedding now (October) in some plants in Thana. Fresh foliage appears generally with the appearance of the flowers or immediately before the flowers are out. Some say that fresh foliage appears in March and April; the flowering time is also variously given. Brandis notes that fresh foliage appears in March and April, and the flowers between January and April before the leaves are out. He adds that, in the Konkan, flowers appear in November. In other places I have seen them all throughout the year more or less, but chiefly between January and April. It must be noted, however, that the same plant never flowers and fruits more than twice a year whatever the months it may blossom in. Brandis found it in leaf and flower in December, 1869. Drury says the flowering time is from January to July. Kurz says that the plant flowers in the cold season, and the fruit is ready in the hot season; Brandis notes that the capsules ripen from April onward, and unless pulled off, remain long on the tree, and, I may add, even dehisce *in situ*. Firminger says the fruit is produced in March.

The colour-markings of the flowers are variously given. These, it may be generally observed, must be, in my opinion, attributed to effects of climate and season. Dalzell and Gibson, who are quite familiar with the Konkan Flora, say that the flowers are white. Drury and Hamilton also say that they are white. Brandis says they are white, with yellow dots at the base. Woodrow says that the colour is yellowish-white. Paxton says that the colour of the flowers is yellow (Bot. Dict., p. 378, 1868). Roxburgh refers to a red-flowered species or variety which is found in the vicinity of Malda; its Sanskrit name is *Madhu-Shigru*. I may observe here, however, that *Madhu-Shigru* is not always a red-flowered variety. (*See Raj Nighant*, Mr. M.C. Apte's Edition, *Shloka* 64). The term is also used to represent a white- flowered variety. The red-flowered variety is sweet-scented (*Shloka* 66). Old Indian writers recognize three varieties: *Swet* (the white), *Nil* or *Krishna* (the black), and *Rakta* (the red) (*Nighanta Prakash*). Bhavaprakash mentions two varieties, white and red. Udoy Chund Dutt notes that the root of the white variety is said to be a stronger rubefacient, while that of the red is preferred for internal use for promoting the appetite and acting on the excretions ("Sanskrit Materia Medica", p. 118). In the Benares Edition of Raj Nighant (1883, p. 58) four varieties of Shegat are given, viz. — yellow, black, white and red. There is really no black variety. By the term *Nil* or *Krishna* it is meant to express probably the greenish colour which the petaloid sepals often assume.

With regard to the fruit, it may be said that it is useful to the natives of this country as a vegetable. The unripe capsules are cooked and eaten with relish. To some Europeans it affords a good substitute for asparagus, and I can say from personal experience that it is by no means a bad substitute, only that it should be properly cooked. Besides, it must be remembered that the fruit of different trees is not alike in taste. Some capsules are distinctly sweet, others are bitterish. It is the sweet variety which has the taste and flavour of asparagus. Not unoften the taste of some capsules is so bitter as to render a seasoned dish void of all enjoyment. The leaves and flowers are often cooked as vegetables in fritters and curries, but they have a mawkish taste; the leaves are more so than the flowers.

With respect to the gum which exudes from the bark of the plant, it may be observed that it is insoluble in water, and has a strong resemblance to that known as tragacanth obtained from *Astragalus verus*.

The seeds, observes Dr. Hamilton, appear somewhat like pith-balls when their winged papery envelope is removed. The seeds, upon dividing them with the nail, says he, are found to abound in a clear, colourless, tasteless, scentless oil, of which the proportion is so large that it may be expressed from good fresh seeds by the simple pressure of the nail. O'Shaughnessy says that the seeds contain so much oil that it exudes on compression between two fingers. "Geoffry informs us," says Hamilton, "that he obtained $30\frac{1}{2}$ ounces of oil from 8 pounds of the decorticated seeds, being at the rate of very nearly 24 lbs. of oil from

100 lbs. of seed." The oil thus obtained is the celebrated oil of Ben or Behen, which at one period constituted a valuable branch of commerce with the East, until excessive imposts and extensive adulteration brought it into unmerited disrepute. Dr. Hamilton distinctly calls *Moringa pterygosperma* the "Oil of Ben" tree. In his Catalogue of the Economic products of the Bombay Presidency (p. 149, 1862), Sir George Birdwood states, what is well-known here-in the Bombay Presidency at least, — that within his observation no oil is obtained from the seeds of *Moringa pterygosperma.* The people use the fruit so largely as a vegetable when it is young and tender that it is removed from the tree long before the seed matures. The best fruit for culinary purposes is one in which the seed has but just formed. From what Mukherji says, it appears that what is called oil of Ben is derived from other seed than those of *Moringa pterygosperma*; for he says that the seeds of *Moringa pterygosperma* yield an oil "*similar to the oil of Ben.*" Not only does Hamilton say that *M. pterygosperma* is the source of Ben oil, but even Lindley, Henfrey, Drury, Dalzell and Gibson say the same thing. Balfour and others say that *M. pterygosperma* produces Oil of Ben. Balfour moreover adds that the oil is an aperient and used by the native doctors in gout and rheumatism. Brandis Baillon, Le Maout and Decaisne, on the other hand, differ from the writers just mentioned on this point. They say that Ben Oil, as it used to be sold in the market in former days, was derived from the true Ben-nuts which are the seeds of *M. aptera.* This is what Brandis says regarding the subject — "From the seeds of another species with *unwinged* seeds, *M. aptera.*, Gaertn, of Africa, a valuable oil (the Ben oil of watchmakers and jewellers) is obtained; but so far as is known, no oil is extracted from the seeds of this species in India." The following are the observations made by Baillon on this point — "The oil of Ben (*M. aptera*) is sweet, inodorous and limpid; does not easily turn rancid, and finally separates into two parts, one of which coagulates (in cold countries observe. - K.R.K.), while the other always remains fluid. It is used in the extraction of certain perfumes, and the more fluid part is often used by the watch-maker." Probably oil might be extracted, says Baillon, from the winged Ben from tropical India, which also contains a quantity of oil, but it does not appear to have been much used hitherto. Le Maout and Decaisne

observes that the *Moringeae* are tropical Asiatic, Arabian and Madagascar trees. The best- known species is *M. aptera*, of which the seed called Ben-nut yields a fixed oil of much repute in the East, because it does not become rancid and is used by perfumers and mechanists in cold countries from its not freezing.

The concluding remarks of Dr. Hamilton's paper before the Pharmaceutical Society I have been hitherto quoting are so graphic, and seem to me to be so useful to lovers and manufactures of scents, that I am tempted to quote them *in extenso*, in the hope that they may encourage some of our indigenous scent-manufactures to cultivate the *Moringa pterygosperma* plant for its seed, with the view of obtaining from the seed that bland oil which has been in the past a very useful scent-storer, and may in future add largely to the manufacture of numerous essential oils and essences from indigenous scented flowers and leaves. To quote, then, the remarks of Dr. Hamilton, he says — "The oil which is so profusely obtained from the seeds is peculiarly valuable for the formation of ointments, from its capability of being kept for almost any length of time without entering into combination with oxygen.

"This property, together with the total absence of colour, smell, and taste, peculiarly adapts it to the purposes of the perfumer, who is able to make it the medium of arrestng the flight of those highly volatile particles of essential oil which constitute the aroma of many of the most odoriferous flowers, and cannot be obtained, by any other means, in a concentrated and permanent form. To effect this, the petals of the flowers, whose odour it is desired to obtain, are thinly spread over flakes of cotton-wool saturated with this oil and the whole inclosed in air-tight tin cases where they are replaced by fresh ones, and the process thus continued till the oil has absorbed so much as was desired of the aroma; it is then separated from the wool by pressure, and preserved under the name of *essence* in well-stoppered bottles. By digesting the oil thus impregnated in alcohol, which does not take up the fixed oil, a solution of the aroma is effected in the spirit, and many odoriferous tinctures or waters, as they are somewhat inaccurately termed, prepared which could not otherwise be obtained. By this process most delicious perfumes might be obtained from the flowers of the *Acacia tortuosa, Pancratium carribeum, Plumeria alba, Plumeria rubra*, & innumerable other flowers of the most exquisite fragrance, which abound within the tropics, blooming unregarded, and wasting their odours on the barren air." It would hardly be within the scope of these remarks to add to this interesting quotation the names of such of our odoriferous plants as would give really very agreeable and delicate oils and essences, and open up a wide field of researche and occupation to our would-be scent-growers were it possible for them to obtain a sufficinet quantity of the oil of the *Moringa pterygosperma* seed. As it is my intention to write an independent paper on the scent-producing plants of Western India on a future occasion, I do not pursue the subject here any further.

The following notes from Watt's Chemical Dictionary may be read usefully in connection with the foregoing observations —Benic acid, obtained from oil of Ben, and sillistearic acid, from the fruit *Stillingia sebifera*, have the same composition (Watt's Chemical Dictionary, Vol. I, page 538). Oil of Ben (Watt's Chemical Dictionary, Vol. I, page 538) constitutes the oil expressed from the fruits of *Moringa Nux Behen*, Desf.; *Guilandina moringa*, Linn.; or *Moringa oleifra*, Lam. It is colourless or slightly yellow, of specific gravity 0.912, thick at 15 C, solid in winter.* It has a mild taste, is neutral to test paper, and does not readily become rancid. "It is used in perfumery to extract the odorous principles of fragrant plants. In India it is used as an embrocation for rheumatism. According to Volcker (Ann. Chem. Pharm., LXIV, 342), it is saponified perfectly by potash, 400 grammes yielding about 18 grammes of a mixture of soild fatty acids together with oleic acid. The soild fatty acids are — 1. An acid soluble in strong alcohol, insoluble in ordinary alcohol, melting at 83 C, and containing 81.6 per cent. carbon and 13.8 hydrogen, numbers which approximate the formula $C_{45} H_{90} O_2$; but the quantity obtained was too small for complete investigation†. 2. Ordinary margaric acid. 3. An acid resembling stearic acid and called by Volcker *Benic acid*."

* *N.B.* - In Europe. - K.R.K.

† Here is a field for further analytical research – K.R.K.

"Another kind of oil of Ben, said to be obtained from the seeds of *Morigna aptera* yields by saponification four fixed fatty acids, *viz.*, stearic acid, margaric acid, and two peculiar acids, *benic acid** and *moringic acid* (Walter, Compt. Rend. XXII, 1143)."

With regard to the fragile nature of the young branches of this tree, that late Mr. Govindji Narayan, who spent his valuable life in enriching the Marathi literature with original, entertaining and instructive works, and whose graphic descriptive power no Marathi writer of his day, except the renowned Rao Sahed Mahadeo Govind Shastri Kolhatkar, ever, surpassed, if any equalled, makes the following remarks in his *Vriksh Varnan* (Description of Plants) — "The *Shevga* tree is so fragile that even by an ordinary violent wind its branches break. The tree is sometimes so over-loaded with capsules that the branches break even by the very weight of the fruit." Mr. Govindji also notes that the trees which grow in the Bassein Prant afford fruit which is excellent eating; that they grow in any kind of soil, and do not require any special care. He corroborates my observation that the trees flower all throughout the year. He describes the flowers as being white. Another indigenous writer, who appears to be well informed in Indian Plants, Raghunathji Indraji of Junagadh, who is popularly known as Kata Bhat, in his *Nighant Sungraha* (1893, p. 320) makes the following observations regarding the Shegat plant — "The flowers are white, red, or greenish. The white-flowered variety is the most common. Its green bark is very pungent, and has the smell of powdered mustard. The juice of the bark yields an oil. Both the juice and the oil are used externally to reduce swellings and enlarged glands." Mr. Vasudev C. Bapat, another writer, a resident of the Ratnagiri District, recognizes only one variety of *Shevga*. (*Vanaushadhi prakash*, Vol. II, page 419).

Moringa pterygosperma as a member of the phanerogamia is a plant of very great interest. In the classification of plants it has been difficult for botanists to assign this plant its proper place. It has ever been a puzzle under what *natural order* to class it. In the days

*N.B. – Benic acid is a name applied to two different fatty acids, both obtained from oil of ben, the one by Volcker, the other by Walter. For distanction, Volcker's acid, which has the higher melting point, may be called the benosteraic acid and Walter's benomargaric acid.

of my boyhood I used to hear of the fifth proposition in the first book of Euclid as the *Pons asinorum.* Were it pardonable to parody this quaint expression, in the advanced days of my scholastic career, I should say that the *Moringa pterygosperma* has ever been the *Pons botanicorum praestantissimorum.* For witness how various botanists have variously attempted to classify this plant, and yet how far we are, even at the present day, from arriving at a satisfactory solution of the difficult question of giving it its proper place among the natural orders.

The classification of plants in *Hooker's flora of British India,* which, from the extreme care with which it is compiled, with the assistance of some of the best and most accomplished botanists now living, must at all times, at any rate for many years to come, be a standard work of reference to Indian botanists, is based on Bentham and Hooker's exhaustive *Genera Plantarum.* Hooker accordingly places the Natural order *Moringeae* next to *Sapindaceae* among the *Disciflorae,* a division which was generally, in days gone by included among the *Thalamiflorae,* but which was first made into a separate division by Bentham and Hooker. The following are the characteristics of the division *Disciflorae* — "Flowers generally with a conspicuous disk, on or about which the stamens, which are nearly always definite, are inserted. The flowers are mostly regular; in the *Moringeae* they are markedly irregular, somewhat resembling those of the *Papilionaceae,* which form one of the sub-orders of the natural order *Leguminosae.*

"*The Moringa pterygosperma* tree" says Mr. E.M. Holmes, F.L.S. (*The Tropical Agriculturist of Ceylon,* Vol. III, 1883-84, page, 285), "is a very interesting one from a botanical point of view, veing allied to the *Leguminosae* in habit, and, indeed, very erroneously included by Linnaeus in the genus *Guilandina.* It resembles the plants of this family in having compound leaves, stipules" (glands according to some botanists, as already mentioned. — K.R.K.) "and flowers which chiefly differ from those of the tribe *Caesalpineae* in the odd petal being inferior, in the one-celled anthers, tricapillary ovary, and anatropous ovules. In the last two characters it approaches *Violaceae,* as well as in the three-valved fruit, parietal placentation, and hollow apex of the style. In properties it resembles the *Cruciferae, Capparidaceae* and *Resedaceae.* By Grisebach it was placed in the *Capparidaceae*, and by other botanists it has been compared with the *Polygalaceae*, *Bignonaceae*, and *Sapindaceae*." Mr. Holmes further remarks that "in the classical General Plantarum of Bentham and Hooker, it follows *Sapindaceae as an anomalous genus of doubtful affinity*."

Lindley has written an elaborate and learned note on the *Moringaceae* or "Moringads" as he calls them. He classes them among the *Hypogynous exogens, Violales*. "This is a little group of small trees," says he, "with an appearance so peculiar that one hardly knows with what to compare them. It seems, however, generally admitted that they resemble some plants of the *Leguminosae*, and it is to the vicinity of those that all botanists, except myself, seem agreed in referring them, moved thereto by their pinnated leaves, with glands between the leaflets, declinate decandrous perigynous stamens, and pod-like fruit. DeCandolle, who did not overlook the anomalous structure of *Moringa* as a Leguminous plant, accounted for the compound nature of its fruit on the supposition that although unity of carpels is the normal structure of the *leguminosae*, yet the presence of more ovaries than one in a few instances of that order explained the constantly trilocular state of that of *Moringa*." I may observe here that the term unity of carpels used by Lindley is likely to mislead not a few. What Lindley means is "one-ness," "single-celled" nature and not union of many carpels or "cohesion;" for let it be remembered that unity also means "union;" witness the old adage "unity is strength." It would be well to briefly state here what these leguminous plants bearing a many-celled fruit are. Loudon says that *Callistachys lanceolata* and *Callistachys ovata*, belonging to the *Leguminosae* bear pods, which are many-celled, stalked, and woody before ripening.* They are both natives of Australia; very handsome shrubs bearing spikes of yellow flowers. Another leguminous plant, *Afzelia africana*, a tree growing in the Colony of Sierra Leone, and bearing black seeds with a scarlet arillus, has many-celled pods†. "It was named by Sir J.E. Smith," says Loudon, "after Dr. Adam Afzelius, an amiable and excellent Swedish botanist, resident,

* *Encyclopædia of Plants*, p. 338, 1829

† *Op. cit.*, p. 339, 1829

for many years in the service of the African Company, at the Colony of Sierra Leone."* *Cathartocarpus Javanicus*, a native of Java, as the name indicates, is another of the leguminous plants bearing pink flowers, which has a many -celled pod†. *Cassia fistula*, our well-known *Bahawa* palnt, so familar in the Konkan and strikingly beautiful, with its dense, long, bright, yellow panicles, reminding one of the poet's "Laburnum drooping gold," bears a long woody pod, which botanists are accustomed to describe as spuriously many-celled. The spurious dissepiments in this pod, be it noted, are horizontal and regularly arranged between each seed as it lies imbedded in a black honey-sweet pulp. Such are the deviations of the fruit among the *Leguminosae* from the normal one-celled ovary.

To resume our reference to Lindley, "it has, however, always seemed to me," says he, "that the resemblance which botanists have found with leguminous order, are" (? is — K.R.K.) "trifling, while the discrepancies are of the first importance. For example, the habit of the plant consists of aa doubly pinnated foliage, which would do as well for Roseworts, or Citronworts, or Rueworts‡; the declinate stamens are found in Rueworts, Milkworts, Capparids, and many others; and as to the pod-like form of the fruit, it is not worth a thought. The objections are that the sepals are of the same texture as the petals, the anthers one-celled, the ovary composed of three carpels which have not the power of turning inward their sides so as to form dissepiments, and that the attachment of the carpel is strictly parietal. It is true that the latter circumstance will not be so much at variance with the leguminous structure as it appears to be, if it should be proved that the sutural and parietal placentation are of the same nature, which seems to be the fact. But connected as it is with other points of difference, and considering that it is parietal placentation in excess, it appears to be of considerable moment. This has always led me to regard the Moringads as members of some great parietal alliance and as claimants of a nearer affinity

* *Op. cit* p. 347, 1829

† *Op. cit* p. 339, 1829

‡ Note that both the Citronworts and Rueworts of Lindley are now classed under one head called the *Rutaceae* - (K.R.K.)

with Violetworts* (*Violales*) than with any other order. And to this opinion I adhere for the following reasons The stamens are definite in number, the corolla is manifestly irregular, the placentation is parietal, and the flowers are not isomeric, the parts of fruit being three, while those of the calyx, corolla, and stamens are five. The main objection to this view is derived from the stamens being perigynous, and it will be seen from the altered arrangements introduced into the present volume" (from which these remarks are quoted. — K.R.K.) "that I now attach much more importance than formerly. But it must be remembered that *Moringa* is not at all more perigynous that *Verrucularia* and others among Malpighiads, or than Reseda among the Crucifers, or than Escholtzia among Poppyworts; and that, in fact, it may be very well regarded as standing in the same relation to Violetworts as *Escholtzia* to Poppyworts. While, however, the parietal placentation seems to turn the scale in favour of the near affinity of the Moringads to Violetworts, there can be little doubt that they also approach the anisomerous Sapindal alliance, especially Milkworts, in their declinate stamens, one-celled anthers and petaloid calyx." Such are the learned observations of the accomplished Dr. Lindley of unrivalled botanic fame. To question his authority would be mere presumption.

The following observation in Professor Henfrey's Botany (edited by Masters and Bennett, p.235, 1884) may be read usefully in connection with the foregoing remarks of Lindley. "The species of the order *Moringaceae* have been referred to the vicinity of the *Leguminosae* principally on account of their per'gynous irregualr flowers, pinnate leaves and pod-like fruit. The structure of the ovary entirely removes them widely from the *Leguminosae* on account of the parietal placentation. Since, judging from the *Rosaceae* the occurrence of additional carpels in *Leguminosae* would be accompanied by an apocarpus condition or at least by axile placentation. *Diphaca* and *Caesalpinia digyna* (Leguminosae) are in fact described

* For the proper understanding of Lindley's argument, I may here mention the natural orders which he includes among the Violales. They are as follows :-

Flacourtiaceae, Lacistemaceae, Samydaceae, Passifloraceae, Malsherbaceae, Moringaceae, Violaceae, Frankeniaceae, Tamaricaceae, Sauvagiaceae, Crassulaceae, and Turneraceae.

as having two legumes*; but the monstrous forms of *Gleditschia* referred to by De-Candolle are said to have two coalescent carpels. Hence Lindley places this order in the neighbourhood of *Violaceae* and conceives it approaches *Polygalaceae*. Others place it between *Capparids* and *Resedaceae*, to the former of which it is closely allied."

Passing on to another botanist, although Baillon includes the *Moringa* series in the nat. ord. *Capparidacae*, he begins his remarks by saying that the series represented by the Moringa is of somewhat doubtful position among the *Capparidaceae*. Baillon is of opinion that *Moringa* shows close analogies with *Leguminosae*, differing essentially only in the number of its parietal placentas; he says that *Cordyla*, which belongs to the leguminous series *Tounateae* or *Swartzieae*, was formerly placed in Capparidaceae, and is only distinguished therefrom by its single placenta. Baillon gives a reference to *Adansonia* (IX, 335) for the enumeration of the different groups to which *Moringa* has been referred, even bringing it near *Monopetace*, such as *Bignoniaceae*. I am sorry I have not the means of referring to Adansonia, else I should have been able to make this part of my paper complete. One has to be content in this country with only a few books of reference in the absence of a well-furnished scientific library, and make the most of scanty resources for acquiring scientific knowledge. According to J.G. Agardh, the *Moringaceae*, though collateral with the *Capparidaceae*, appear to be more allied to the *Leguminosae* on account of their pentamerous, nearly symmetrical flowers, and perigynous stamens.

Le Maout† and Decaisne are more decided in their opinion regarding the position of the *Moringeae* among the natural orders. "It is amongst the *Capparideae*," they say, "that we must search for the real affinities of *Moringeae*." "The genus *Moringa*," they observe, "has been by some botanists placed in *Papilionaceae* on account of a slight resemblance in the flower, which, however, indicates no true

* A double ovary sometimes occurs as a monstrosity in *Wistaria sinensis*, in *Gleditschia* and in the French bean (*Phaseolus*); and a Mimosa with 5 carpels (thus a symmetrical) flower is said to have been seen by St. Hilaire (Henfrey p. 239).

† Descriptive Botany (translated by Mrs. Hooker), p. 236, 1873.

affinity. Hooker (following Lindley) compared it with *Violarieae*, "which resemble it in their irregular flower, with unequal dorsal petal, in their perigynous insertion, tubular style, one-celled ovary with three parietal nerviform placentas and anatropous ovules; but *Moringeae* are widely separated by habit, one-celled anthers and exalbuminous seeds." The affinities of *Moringeae* with the *Capparideae* are to be observed "in their polypetalous imbricate corolla, perigynism, stamens more numerous than petals, stipitate one-celled ovary, parietal placentation, siliquose capsule, exalbuminous embryo, alternate leaves, and caducous stipules." To these must be added *the acrid root, leaves*, and *bark* which are common to both families, recalling the smell and taste of the horse-radish, and associating Moringa with the *Cruciferae*, themselves so closely allied to *Capparideae*.

Kurz places this plant between the *Capparideae* and *Violaceae*, as orders belonging to sub-division *Parietales* (i.e., having parietal placentas) of the thalamiflorous plants. The observations of Wight on this question are too interesting to be omitted in such a general consideration of the true position of the *Moringa* in the series of natural orders. He realizes the difficulty of assigning the *Moringa* its exact place. "This small order," says he,* "consisting of one genus and about three species, long associated with *Leguminoseae* though sufficiently distinct, and so peculiar in its relations that it is not yet distinct where it should stand in the series of orders, the structure of the ovary placing it among the parietose group, near *Violarieae*, while according to the flowers it is more justly referable to the tribe *Caesalpineae* of the *Leguminosae*." Notice that some botanists think the *Moringa* is more allied to the sub-order *Papilionaceae*. "The characters derived from the tribe *Caesalpineae* of *Leguminosae* as being the more conspicuous of the two seem to have induced botanists to prefer placing it near the *Leguminosae*, the more so as the habit still further associates it with that order."

* * * * * * * * *

"*Affinities*. — As stated above, these are very obscure owing to the peculiar structure of the ovary, one-celled with three parietal placentae, the fruit afterwards spliting into three valves. Arnott (*Encylopaedia*

* *Wight's Illustrations of Botany*, pp. 186-187, 1840.

Britannica, 7th Edition) remarks, formerly considered as part of the *Leguminosae*, but not separated by Mr. Brown. It seems, however, to have more affinity with these than with others; nor do we see in the fruit very grave objections to such a supposition, as the flowers of *Gleditschia* have occasionally two carpels united in the same manner as the three of Moringa."

"Of the correctness of this view, or otherwise," Wight goes on to say, "I can offer no opinion, being altogether unacquainted with *Gleditschia*'* Lindley however holds on opposite opinion; he says, 'DeCandolle, who did not overlook its anomalous structure as a leguminous plant, accounted for the compound natue of its fruit upon the supposition that though *unity of carpels* is the normal structure of *Leguminoseae*, yet the presence of more ovaries than one in a few instances in that order explained the constantly trilocular state of that of *Moringa*. To this, however, there are numerous and grave objections which cannot fail to strike every botanist. What its proper station should be remains to be determined;' again, Decaisne seems to think it has more affinity with *Leguminoseae* than with any other order. On these passages it may be observed," says Wight, "that it would have been more satisfactory had some of the grave objctions been stated, since it is quite undeniable that Messrs. DeCandolle, Arnott, and Decaisne are all eminent botanists and yet they do not seem to have been struck by them. While I thus object to such sweeping conclusions as the above, I confess I cannot help coinciding with Dr. Lindley in thinking that the doctrine he opposes is pushed, in this instance, much too far by these authors, though I scarcely feel myslef sufficiently master of the subject to join issue with them on this, one of the most difficult questions in botany."

I have thus placed before my readers the different view expressed by eminent botanists regarding the position of the *Moringa* plant in the series of natural orders; and must now bring to a close the lengthy quotation, which I have introduced more for the purpose of bringing together the scattered information on this point than with any hope of arriving at a definite conclusion as to which of the

* A north American genus, of which the honey-locust tree is a species (*G. triacanthos*). There are two chinese species of this genus besides.

natural orders *Moringa pterygosperma* should finally belong. In monographic papers of the sort I am writing in this series, it seems desirable that all that is worth knowing should be stringed together; that is my only apology for such lengthy extracts.

I might add a distinct chapter describing the various savoury Indian dishes that could be turned out of the fruit, flower, and leaf of *Moringa*, with the requisite flavouring ingredients, such as pepper (red and black), salt, cocoanut scrapings, cumin seed, onion, &c., but it would be foreign to the purposes of the paper to do so. In fact, knowing how valuable the whole of this plant is from a culinary point of veiw, and judging from my own personal taste and experience what a substantial article of diet the *Moringa* capsule is, I am conscious I shall be asked why I include this tree among the poisonous plants of Bombay. Not only will the captious critic, but even the earnest inquirer, the careful student in search of botanical knowledge, ask me to explain what is apparently a paradoxical statement, *viz.*, that the plant is both poisonous and wholesome. This I shall try to explain in the following remarks.

THE POISONOUS PROPERTIES

It is not by any means to be supposed that, because one part of a plant is edible, any other part of the same plant is not poisonus, using the word poison in its broadest sense to signify whatever is hurtful to the human body. Instances are not unknown among other plants, where one part of the same is safely used for culinary purposes, whereas another has proved to be distinctly poisonous. French beans, for instance, are known to serve as an excellent vegetable. Few will believe, however, that the root of the French bean runner when eaten is poisonous. Mr. Bartlett has recorded an instance (*Pharmaceutical Journal*, Vol. II, page 721, 1842-43), in which the poisonous effects of the root were distinctly noted. "A customer of mine," says he, "last Christmas was digging up some roots of the French beans or Scarlet runners and, observing them to be very large and look very white when broken, thought he would taste them, thinking there could be no harm in partaking of the root of a vegetable which had formely supplied his table. His family and maid-servant being also in the garden, all partook of some, pronouncing them good and pleasant to the taste; presently, the servant complained of giddiness and sickness, and while rendering assistance, the master himself, with the mistress and five children, followed in the train; the sickness continuing for half an hour."

Another instance occurred last year in my own practice in the Thana Jail. For years it was the practice in the Thana Jail to cook up the tender tops and leaves of the sweet potato (*Batatas edulis*) and serve as vegetables to the jail inmates without any deleterious effects. One night, last cold weather, I was suddenly called by my Jail Hospital Assistant to attend to five or six cases of almost epidemic vomiting and purging, in a more or less severe form, among prisoners both male and female in the different barracks of the jail. These five or six cases rose to fifty within the next two hours. If I had been hasty in my investigation of these cases, I might have put down this sudden outburst of vomiting and purging to an epidemic of cholera. But there was no cholera in the town, nor was it raging in the districts around, which supply our jail with its daily fluctuating population. I made a careful examination of the vomits of the affected prisoners, and in those vomits I found undigested portions of the sweet potato leaves, which I though might be the probable cause of this almost epidemic suffering of the jail inmates, the jail population at the time being over 700 souls all told. To have to treat fifty prisoners within two hours is decidedly a serious affair. Under prompt treatment, the details of which need not be mentioned here, fortunately for all concerned, there was no death. My next step was to obtain a sufficient quantity of the leaves and tender tops of the sweet- potato plant, which I submitted for analysis to my freind Mr. Stephenson, F.C.S., then of Messrs. Kemp & Co. Mr. Stephenson discovered a glucoside in the plant which I consider was the cause of the ailment. It is well known that at certain times plants, which produce substantial roots or tubers, are rich in glucosides in the sap of their leaves before they finally store up all their starch in their roots and tubers. The glucoside discovered by Mr. Stephenson was probably not unlike that found in the tubers of Jalap (*Exogonium purga*) belonging to the nat. ord. *Convolvulaceae*, of which the sweet-potato is no mean member.

There are several other instances which may be cited here to show that even although one part of a plant may be perfectly innocuous, another may be obnoxious. Thus, *Abrus precatorius* (Marathi *Gunj*, nat. ord. *Leguminosae*), the root of which is perfectly harmless and considered a safe substitute for the liquorice-root (*Glycerrhiza glabra*, nat. ord. *Leguminosae*), bears seeds which, if used internally, act as powerful gastric irritants producing symptoms not easily to be differentiated from those of cholera. I have several instances of such affection, which I shall mention in detail when I describe the plant in a future number of this series of papers on the poisonous plants of Bombay.

Take, again, the mature seeds of *Badyan* (*Illicium anisatum*, nat. ord. *Magnoliaceae*). The curious looking star- like fruit of this tree is largely employed as a carminative, and serves as a valuable and much appreciated flavouring adjunct in Indian curries and *pullaos*. But the seed is distinctly poisonous if eaten when perfectly mature. Dr. Rozario of Mazagon has come across some cases of poisoning from such seeds. In my own practice at the J.J. Hospital of Bombay, I came across a case in 1887. Mr. J. Gallagher, the House Surgeon then, brought the case to my notice. He has since recorded another case of a fatal nature which occurred very recently in the J.J. Hospital, where the seeds had been swallowed outside the hospital.

One more instance I may add to these. From personal experience I know that a quarter-inch square piece of the bark of *Mimusops elengi* (Marathi *bakul* or *Wowlen*, nat. ord. *Sapotaceae*) chewed at the instance of a zealous benefactor for relieving a bad tooth-ache produced within twelve hours a terrible oedema of the uvula and the mucous membrane of the whole mouth, and an incessant flow of saliva for over two hours. Fortunately the bark was only chewed and its irritant juice not swallowed, else to a certainty violent gastric inflammation had followed. This was a case of the patient ejaculating "*Save me from my friends!*" Be it noted, on the other hand, that the fruit of Bakul is edible and perfectly harmless. It has a sweet astringent taste.

I have thus tried to justify my insertion of the *Moringa pterygosperma* plant among the poisonous plants, however largely and without harm it is used for culinary purposes.

Now let me turn to the consideration of what constitutes the poisonous nature of the plant. Rheede was not unacquainted with the poisonous nature of the root of this plant. His knowledge on this score was mainly derived from the native Vaidyas, learned in the existing flora of the Malabar tract of this country, who materially helped him in the production of that encyclopaediac work, known as the *Hortus Malabaricus*. "The root," says he, "causes abortion." I may be permitted to add that the bark is equally mischievous. Last year, I had to give evidence in an abortion case before Mr. Patwardhan, the District Deputy Collector and First-Class Magistrate of Thana. In this case, the women, who had successfully used the bark, admitted she had done so. From 2 to 3 lbs. of the pounded bark had been used externally on the hypogastric and public regions for the distinct purpose of procuring abortion. The woman was a poor unfortunate middle aged widow-and the fate of poor helpless young, or middle-aged widows is very hard in this country; she had been deceived and saddled with the burden of a growing foetus; to avoid disgrace she tried to destroy the foetus, and she succeeded. This is an instance of how the *Moringa* bark is abortive. O'Shaughnessy does not notice it in his *Bengal Dispensatory*. All that he says is that the root reduced to a pulp affords an external irritant. Other writers say that the root acts as a rubefacient and vesicant, and causes great pain. Udoy Chandra Dutt says that the root of the white flowered plant is said to be a stronger rubefacient, while that of the bark is preferred for internal use for promoting the appetite and acting on the excretions. (Sanskrit *Materia Medica*, p. 118, 1877).

The ancient Hindu writers, as for instance the author of *Raj Nighant*, of *Bhavprakash* of *Madanpal-Nighant*,* do not refer to the abortifacient property of any part of the *Shegat* plant. In *Amarkosha*, a very old Sanskrit dictionary, *Shegat* is said to be *pungent*†. No mention of the abortifacient property of this plant is made

* Dr. Bonavia gives the date of this work as 1360 A.D. (The Cultivated Oranges and lemons, p. 215, 1890).

† Dr. Bonavia gives the date as the 8th or 9th century A.D., and observes that, according to Reinand, it appears that Amarkosha was written in the 5th century. (*Op. cit*).

in *Dhanvantary Nighant.* None of the later writers, such as the compilers of *Nighant Ratnakar;* Udoy Chandra Dutt, of Calcutta; Dr. Sakharam Arjun, Dr. Pandurang Gopal, and Mr. Govindji Narayan, of Bombay; Kata Bhat, of Junagadh; and Mr. Wasudeo Chintamon Baput of Ratnagiri — make any allusion to its abortifacient effects. Mr. Jaikisson Indraji, of Prebunder, however, distinctly says that the bark is used to procure abortion. "It is also used," says he, "as a substitute for Laminaria tents for dilating the *os uteri.*" He says that the gum is also abortive. Dr. Dymock says the same thing. So far as I know the gum is inert. It is insoluble in water, and if left in it for some time, swells into a gelatinous mass like gum Tragacanth obtained from *Astragalus verus* (nat. ord. *Leguminosae*). "Rumphius and Loureiro state," says Dr. Dymock, "that the bark is emmenagogue and even abortifacient. In Bengal half-ounce doses of the bark are said to be used to procure abortion." This would point to the fact that the bark is used internally, *i.e.*, taken by the stomach. Dr. Dymock, in corroborating the statement of Jaikisson, adds a remark that "it is difficult to obtain reliable evidence upon a point of this nature. It would be quite possible to use it as a tent to dilate the *os uteri as it is very tough* and swells rapidly when moistened." — (*Pharmacographia Indica*, Vol. I, p. 398). That it swells rapidly it cannot be doubted as the wood is porous. But I doubt if any part of *Shegat* can be said to be tough except the pericarp of the capsule. The plant is essentially fragile, and in no way can it be said to be tough. Dr. Dymock himself says that the wood of the root is soft and porous, and that the bark of it is thick and soft. The parenchymatous cells of the bark "are loaded with globules of essential oil" which is highly pungent and very much like the horse-radish oil or mustard oil in its irritating properties. Dr. Dymock observes, however, that "the odour of *Moringa* oil is distinct from that of mustard and garlic, and more offensive." Dr. Dymock further says that the *Moringa* bark "contains a white crystalline principle answering to the reactions of an alkaloid."

My authority for including this tree among the poisonous plants of Bombay is Dr. Norman Chevers. It is on his authority that

* The italics are mine – K.R.K.

Dr. Lyon also includes this tree among the abortifacients. These papers of mine on the poisonous plants of Bombay are essentially supplementary to Dr. Lyons, standard work on Medical Jurisprudence, which in a great measure may be said to be the originator of these my contributions to the *Journal of the Bombay Natural History Society.*

On the authority of that veteran scholar, Rai Bahadur Kannya Lall Dey, C.I.E., who to his extensive knowledge of the botany of India adds the acumen of a careful observer, Dr. Norman Chevers makes the following remarks regarding the *Moringa* plant* — "A piece of *Sujna bark*" (*Sujna* is the Bengal name for *Shegat.* — K.R.K.) "about half an ounce in weight, is pounded with twenty-one black pepper-corns, made into a paste and swallowed to procure abortion. This is said to be a very dangerous means, as the woman, as a rule, dies with the foetus." I may observe that, as stated before, the external application of the pounded bark has succeeded in expelling the foetus, but has left the mother uninjured and fit to be tried in a court of law for foeticide. Dr. Norman Chevers further observes that "in all the cases where the application of the substance *per vaginam* is required, a good amount of tact and skill is needed on the part of the applier."

Be it noted here that, in communicating this valuable information to Dr. Chevers, Rai Bahadur Kannya Lall Dey observes that he has obtained these facts regarding the various modes of producing abortion in use near Calcutta by paying a woman who is perfectly conversant with the practice. Dr. Norman Chevers further says that "he speaks confidently of the statements of Kannya Lall Dey as *facts*, because he finds them to be perfectly in accordance with nearly all that he had previously learnt on the subject."

To pursue the subject further and enter into the question of the physiological action of the bark on the gravid uterus would be perhaps groping in the dark, inasmuch as what is as yet known about the abortive properties of the bark is more in a general way than as the result of physiological or clinical investigation. In other words, it is not known whether the bark produces tonic or clonic contraction of the uterus. It is known as an emmenagogue; and what is an emmenagogue in small doses, may prove, and has proved, aborti-

* *N. Chever's Medical Jurisprudence*, p. 716, 1870.

facient in large doses. Mr. Holmes distinctly observes that "the decoction of the root or of the bark is given as an emmenagogue; but it is said to be liable to produce abortion." The various parts of the plant are diuretic; and diuretics are in a few cases known to have caused abortion (*Ogston's Medical Jurisprudence*, p. 97, 1878). The leaves are said to be emetic. Excessive vomiting, whether due to pregnancy or to emetics which are not unknown to abortion-producers, causes exhaustion of nerve-force and thereby leads to the expulsion of the foetus. Bapat says that four tolas of the juice of leaves produces bilious vomiting. The juice of the bark is said to be purgative, acting especially on the biliary organ; should it by any chance produce hypercatharsis, indirectly it may be the means of causing abortion. Baillon says that the embryo is bitter and purgative. Royle says that the seeds are aperient. The question is, whether in the case of *Moringa*, abortion is caused by any expellat action on the part of the volatile oil contained in the bark acting on the uterus through the mother's blood. It may be noted that no specieal abortive property is required in the bark when used locally, *i.e.*, to the *os* itself. Such artificial direct violence or injury to the *os* or foetus, even when the substance used is physiologically non-irritant, may produce abortion. Under such circumstances, as a general rule, the healthy uterus takes long before it expels a healthy foetus. Dr. Barnes lays it down as an axiom that "*the healthy ovum clings to the healthy uterus with wonderful tenacity*." It is where the system is diseased, weakened and worn out by incessant worry, anxiety, great mental prostration, starvation, exposure and other weakening causes, that such local means take ready effect.

It is for the future investigator to find out the physiological effects of the volatile oil of *Moringa* on the uterine system. Let it be remembered that the bland oil expressed from the seed of *Moringa* is entirely different from the volatile oil found in the leaves, root and bark, which latter has a disagreeable odour, and is similar to the mustard and horse-radish oils. The oil of the seeds, apart from its use and importance in the manufacture of scents, is non-irritating. Let me quote again the remarks of Holmes on this point. "It is a noteworthy point", says he, "that whereas the whole plant is so offensive, its seeds are scentless to a remarkable and almost to an incredible degree; nevertheless the practical experience of all previous observers goes to show that the seed oil is absolutely scentless, and free from all irritating properties."

Carl Hubatka has shown that the mustard and horse-radish oils are identical in composition (Ph. Journal, p. 291, Vol. III, 1843- 44).*

"It is singular," says he, "that man should have selected horse-radish and mustard as condiments on account of their analogy in taste, as he also thus detected the analogy between tea and coffee." In contrast with this, Broughton says that the oil of the Moringa plant is different from either oil of mustard or oil of garlic. All these oils, however, belong to the Allyl compounds. (*Roscoe and Schorlemmer's Organic Chemistry*, Vol. III, Pt. 2, pp. 436-451). They are not only disagreeable but pungent. Roscoe observes "That the Attic Greek comedians mention mustard as a biting substance which causes a flow of tears, but is an excellent substance for giving a, relish to certain kinds of food." Such is also the experience of those who greedily chew an extra large bit of the scrapings of the *Moringa root* at an Anglo-Indian table. I wish I could say that this flow of tears is not unknown to me personally !

* Quoted from Ann. der.Chem. und Pharm. xlviii, pp. 153-157.

Reprinted from
Journal Bombay Natural History Society
Vol. 9(3): pp. 235-257. 1894

PART XI.

SEMECARPUS ANACARDIUM - (*Linn.*)
Natural Order — ANACARDIACEÆ
MARATHI (BIBÂ); (BHILÂWÂ) बिबा; भिलावा

(Read before the Bombay Natural History Society on 28th January 1895).

This is the marking-nut tree of English writers.

A deciduous forest-tree common in several parts of India and Other tropical countries, varying in height from 20 to 40 feet, very showy when in full blossom and full fruit. Very common in the jungles of Thana. Old leaves shed in January and February. New leaves appear in April and May. The tree blossoms soon after the new leaves are thrown out. The flowering time lasts from May to August. The seed is ripe in January and February. (Roxburgh).

BRANCHES — Numerous, spreading.

TRUNK. — Straight; covered with grey, scabrous, bark, which is often brown. Girth 4 to 6 feet.

BARK. — "1 inch thick, dusky grey, blackish with irregularly quadrangular plates, separated by narrow longitudinal and shallow, irregular cross furrows or wrinkles, in old trees rough with exfoliating scales." (Brandis' Forest Flora of N.W. and C.India, page 125). "The inner substance of the bark contains," says Roxburgh, "in crevices, a quantity of *white,** soft, almost insipid gum." He further adds that "from wounds made in the bark, a dirty looking *brownish** soft gum is procured, which dissolves slowly in the mouth." The latter remark is nearer the truth, according to my experience. The resinous exudation from the young fruit, however, is milk-white or cream-coloured as it escapes; it soon turns jet-black on exposure.

THE WOOD. — says Brandis, "is ash-coloured, reddish-white or brown, even but open-grained." It is full of acrid juice which causes swelling and irritation. "Timber-cutters," says Brandis, "object to felling it unless it has been ringed for some time." It cracks in seasoning; is not durable and not much used. Weight, 42 lbs. per cubic foot. Sap-wood and heart-wood not distinct (Brandis). I find that under an ordinary magnifying glass the medullary rays are fairly well distinct. Kurz observes that the wood is grey or reddish-white, soft and useless. (Forest Flora of Burma, Vol. I, pp. 311-312.

LEAVES. — Generally closely arranged at the extremities of the branchlets; simple, alternate, very coriaceous and flat; very large, 9 to 30 inches long, 5 to 12 inches broad; cuneate, oblong or obovate-oblong, rarely linear-oblong. Margin quite entire; cartilaginous; usually contracted below the middle, not unoften also above the middle, giving the whole leaf a wavy appearance slightly. Apex rounded, "blunt or mucronate," says Kurz. Base rounded, cordate or cuneate. The surface is opaque above, and slightly pubescent, especially when young; whitish or glaucous and thickly pubescent beneath. Hooker says that the leaves are brown beneath. This is not the case in the Konkan at any rate. Kurz observes that the leaves of the shoots or young trees are not seldom elongate, and cuneate obovate with a sharp point. *Nerves* 16 to 25 pairs; stout, slightly arched, pale white. The entire net-venation is boldly prominent and very coarse, especially on the under-surface, giving the leaves a notably harsh feel. It is difficult to examine the

* The italics are mine – K.R.K.

stomata under the microscope on account of this prominent net-work of veins or nerves. Kurz agrees with me in the view that the venation is markedly coarse on the under-surface. Dalzell and Gibson, who were particularly familiar with the Bombay specimens also say that the leaves are whitish and tomentose beneath.

PETIOLE. — Short, varying in length from 1 to 2 inches, rounded or half-rounded, stout, not winged, densely puberulous.

FLOWERS. — Polygamous or dioecious; small, $\frac{1}{4}$ to $\frac{1}{3}$ inch in diameter; subsessile, fascicled; the fascicles arranged in erect compound terminal panicles.

INFLORESCENCE — In large terminal tomentose panicles; the lower branches of the panicles issuing from the axils of terminal leaves. The panicles are stout, equalling or shorter than the leaves, branching into numerous spikes. The panicles of the male tree much more slender.

BRACTS. — Many, small, lanceolate, fugacious.

BRACTEOLES — Present but fugacious.

ÆSTIVATION — Imbricate.

The above is a description of the flowers generally. It would be useful to note the differences in the male, female and hermaphrodite flowers individually.

FEMALE AND HERMAPHRODITE FLOWERS — says Kurz, "are 1 to 3 lines long, racemulose and nearly twice longer than the clustered, almost sessile, male flowers. Pedicels very thick and tomentose." He further adds that both male and female flowers form a terminal ample robust tomentose panicle of the length of the leaves or shorter.

CALYX. — 5-fid; segments deciduous.

COROLLA. — Greenish-white, or "greenish-yellow." (Brandis).

PETALS. — Five; "valvate-imbricate" (Kurz). 3 to 4 times the length of the calyx; oblong, slightly pointed at the apex, inserted under the margin of the disk; sessile, glabrous, very spreading.

DISK. — Broad, annular, between stamens and ovary.

STAMENS. — 5, alternate with the petals; inserted on the margin of the disk; imperfect or sterile in female flowers; equal; distinct.

FILAMENTS — Subulate from a somewhat dilated base; of the length of the petals.

ANTHERS. — Yellow; ovoid or elliptical.

OVARY. — Free, sessile, 1-celld. "Densely appressed, tawny, hispid," says Kurz.

STYLES — Three, from the apex of the ovary; divergent incrassate.

STIGMA. — Sub-clavate, shortly two-lobed or retuse.

OVULES. — Inserted at the apex of the cell; pendulous from a basal funicle.

MALE FLOWERS — Usually smaller than the hermaphrodite; often on a separate tree. Calyx and corolla as in the hermaphrodite flower. Petals "about a line long; oblong-lanceolate" (Kurz).

FILAMENTS — Five, of the length of the petals.

ANTHERS. — Much larger than in the hermaphrodites.

PISTIL — Absent, or small, and abortive, and in the form of a semi- globular hairy glandular body (Roxburgh).

FRUIT. — A drupe, 1 inch long and about as broad, often less; ovoid, obliquely ovoid or cordate-ovate, with a slight obtuse notch on either side under the apex; unequally compressed; slightly convex in some parts, and quite plain in others; seated at the summit of, or more or less immersed in a fleshy receptacle formed of the thickened disk and accrescent calyx base termed the *Hypocarp.*

CUP — Fleshy, orange-red, smooth, about the size of the nut, sometimes a little larger.

PERICARP — Smooth, shining-black, thick; containing between the outer and inner laminae roundish or oblong cells full of corrosive resinous juice. This juice is white when the fruit is young, darkening on exposure to the air. In the mature fruit it is brownish or perfectly black. The inner lamina is hard and rugose; the outer smooth, leathery and less hard.

SEED — Pendulous; "with a swollen or umbilicate funicle." (Lubbock).

TESTA. — Coriaceous; inner coat somewhat fleshy.

EMBRYO. — Thick, milk-white. Plumule "ovate-leaves, veined, conduplicate, very thin." (Gaertner.)

COTYLEDONS. — Fleshy, thick, white, irregularly plano-convex.

ALBUMEN — Absent.

RADICLE — Superior, minute, concealed within the apex of the cotyledons; "always directed to the hilum." (Lindley's Vegetable Kingdom, p. 465, 1847).

REMARKS

I have said on the authority of Hooker and Kurz that this tree is deciduous. I can also add my own humble testimony to that effect. In Paxton's Botanical Dictionary, however, and in Johnson's Gardener's Dictionary, the plant is said to be an evergreen. I presume that this description is taken from the stove-plants of the marking-nut that are reared in the English nurseries. I think it requires the tropical heat and open air jungle-life to give it its full character of a deciduous tree. Balfour observes that the natural order *Anacardiaceae* is unknown in Australia (Manual of Botany, p. 473, 1875). Hooker, however, notes that the *Semecarpus anacardium* is to be found in North Australia. Baillon also says that the representatives of the genus *Semecarpus* are to be seen in the Oceania, which of course includes Australia. Among introduced plants belonging to the Natural Order *Anacardiaceae* I have seen the mango grow in the beautiful Government Gardens of Sydney. There are several species of the genus *Rhus* (Sumach) also growing there. The mango does not evidently thrive there, and bears very poor fruit, if at all. But several of the American species of *Rhus* seemed to me to be in very good condition, as if born to the soil. The northern part of Australia would not be unnatural soil for the growth of the different species and genera of the Natural Order *Anacardiaceae*. It is therefore, difficult to understand how Professor Balfour made such a wholesale statement as to say that the Natural Order *Anacardiaceae* is unknown in Australia. Perhaps the information we now possess was not available when he wrote his work.

The following are the synonyms as given in Part I of the *Index Kewensis* compiled by B. Daydon Jackson at the expense of the late Charles Robert Darwin of sacred memory, under the direction of Sir Joseph Hooker (p. 115)—

Anacardium latifolium. — Lam. Encyc., I., 139 (Ill. t. 208).

A. longifolium. — Lam. l.c.

A. officinale. — Pritz. Ind. Ic., 59.

A. orientale. — Auct. Ex Steud. Nom., Ed. II., i. 82.

A. officinarum. — Gaertn. Fruct., i. 192, t. 40.

A. solitarium. — Stokes, Bot. Mat. Med., II. 159.

In "Johnson's Gardener's Dictionary" edited by Wight and Dewar (1894 p. 892), *Cassuvium longifolium* is also mentioned as a synonym of *Semecarpus anacardium.* Referring to the *Index Kewensis* under the Head *Cassuvium* I find that this last synonym is not mentioned. It appears, however, that according to Rottboll *Cassuvium* and *Anacardium* are synonymous.

Sir Joseph Hooker also mentions *Semecarpus latifolius* Pers. (ex Marchand), as a synonym of *Semecarpus anacardium.* (Flora Br. Ind., Vol. II. p. 31).

With regard to the habitat of the plant, Dr. Gregg says that it is a native of the mountainous parts of India. Dalzell and Gibson, who knew the Bombay Flora well, say that it is common in the Dekkan and the Konkan, which latter includes the Waree country (Sawant Wadi). Contrast with this the observation of the Rev. Mr. Nairne. He says : "*Semecarpus anacardium* is said to be common in the Dekkan, Konkan and Gujarat, but I should scarcely call it so" (p. 69, The Flowering Plants of Western India). So far as the Thana district is concerned, it is very commonly seen in the jungles. Mr Govindji Narayan, an able indigenous writer, in his Marathi work entitled "*Vriksh- varnan,*" observes that the *Kharvis* of Damaun and Surat use the marking-nut for nearly every kind of ailment.

In describing the *Semecarpus anacardium* of the Colonies in his *Fragmenta phytographiae Australioe,*" Baron Sirt Ferdinand von Mueller observes that "the tree grows to a height of even 50 ft.; and that the fruit is known as the cashew-nut of the Colonies" (p. 23, Vol. VII, 1869-71). It may be observed that the cashew-nut of this country is the kaju plant (*Anacardium occidentale*).

The *Anacardiaceae* were formerly classed as a sub-division or sub-order, by Kunth, Wight and Arnott, and DeCandolle, under the Natural Order *Terebintaceae* of Jussieu. In the *Genera Plantarum,* the Natural Order *Anacardiaceae,* is synonymous with *Terebinthaceae.* The Anacards," says Lindley, "are distinctly known by the seeds hanging from the end of a thread, which rises up from the base of the carpels, which in general are solitary, or at least quite distinct, and are sometimes, when quite ripe, placed at the end of an excessively enlarged disk as in the cashew-nut itself" — (Vegetable Kingdom, p. 465, 1847). "As an ornamental tree, either in full foliage or before the fall of the leaf," says Surgeon-General Balfour, "the marking-nut tree merits observation." He says that the *mature corolla* and receptacle are fleshy and sweetish-sour. As against this, I may observe that the receptacle is fleshy; and not the *corolla*, but the *Calyx-base*. The corolla is distinctly deciduous; so is the upper part of the calyx where the sepals are distinct. "The marking-nut is well described by the Arabs," says Dr. Dymock in one of his earlier notes on this plant, "as resembling the heart of an animal, the torus representing the auricles and the fruit the ventricles." — (Pharmaceut. Journal, p. 1003, Vol. VIII, 1877-78, 3rd Ser.)

"The bark of the younger parts," says Roxburgh, "is smooth;" but from an examination of the plants growing in and around Thana, I find that the younger parts are invariably covered with a light ash-coloured or brownish pubescence. Brandis fully bears me out in this assertion when he says that "young branches, inflorescence, petiole, and underside of leaves are clothed with a short, somewhat harsh tomentum of simple hairs." The simple hairs, as seen by me under the microscope, magnified about two hundred times, are unicellular. Their shape varies much; some are straight and blunt, or sharp-pointed; others wavy and sharp- pointed. These forms are figured under No.5 in the supplementary Plate M. "The bark is astringent and used in dyeing," says Brandis.

Wight and Arnott say that the under-surface of the leaves is "scarcely downy." This is not the case so far as the Thana plants are concerned. If anything, the under-surface is distinctly downy. The tomentum is apparent even to the naked eye. Kurz agrees with me in this, for he says, "the under-surface is covered with a thick grey velvetty tomentum."

Katâ Bhat of Junagadh says that the flowers are red (p. 285, *Nighant-Sangraha*). All the previous writers are unanimous in describing the colour of the flowers as greenish yellow or whitish-green. The stigma is sometimes tinged brownish.

With regard to the one-celled nature of the ovary in *Semecarpus anacardium* I have this to observe. In the tribe *Spondiae*, of the Natural Order *Anacardiaceae*, the ovary is syncarpous, consisting of 2 to 5 carpels and containing as many cells. From this occurrence of several distinct carpels in one flower among the *Spondiae* the one-celled ovary of the *Semecarpus anacardium* "may be considered," says Sir John Lubbock, "as reduced types." (Seedlings, Vol. I., p. 369).

Baillon observes that the cup-shaped fleshy orange-red enlargement, more or less high round the pericarp, and encricling the base of the drupe, is a peduncular swelling. I think it is something more than that. I contains in addition to the peduncular enlargement, the base of an accrescent calyx. Baillon says that fermented drinks and conserves are made from this very showy cup. It has an agreeably acid taste, and is perfectly harmless. In speaking of the edible nature of the fruit of a congener of *Semecarpus anacardium*, namely, the elm-leaved Sumach of America (*Rhus coriaria*), Mr. Nuttall observes that "the pulp of the drupes of several species affords an agreeable acid similar to that of wood-sorrel, either oxalic or tartaric" (p. 122, Vol. II., "North American Sylva). It would be interesting to know to what particular acid the sour taste of the bright orange cups of the marking-nut is due. Mr. Wasudev Chintaman Bapat of Naringre (Ratnagiri district) says that the orange cups are dried and eaten like figs under the name of *Bipti*. The ripe fruits are strung together on a piece of twine and hung up to dry in the sun before being so used. It is stated in Lindley and Moore's "Treasury of Botany" that the *unripe fruit* is employed for making a kind of ink (Vol. II, p. 1047). I may observe that even the ripe and dried fruit is similarly used, I should even say *principally* used. The dried flat cotyledons are used in this country under the name of *godambi*. Though some consider the taste of the fruit agreeable, Brandis is of opinion that the fruit, when fresh, is "acrid and astringent; roasted, it is said to taste somewhat like roasted apples, and when dry, somewhat like dates." Roxburgh says that before being roasted, the fruit leaves sometimes a painful sensation on the tongue. The roasting is usually done in hot ashes. Masters, who writes the article on *Semecarpus anacardium* in the "Treasury of Botany" just referred to, says that "the seeds called Malacca-beans or Marsh-nuts are eaten, and are said to stimulate the mental powers and especially the memory." I know a case of leprosy complicated with loss of memory, in which the nut was used by a Hindu Vaidya with the special object of curing the latter affection.

I may add here that, although on the authority of Mr. Masters I have ventured to state that the seeds of the marking-nut tree are called Malacca-beans, the plant known in American works as the Malacca-bean plant should not be confounded with the marking-nut tree. The former is the *Avicennia tomentosa*, Jacquin, of the natural Order *Myoporinoe* R. Brown. It is now classed under the Natural Order *Verbenaceae* by Bentham and Hooker, and is also known under the name Linnaeus originally gave it, namely, *Avicennia officinalis*. It grows abundantly as a straggling shrub on the entire Bombay and Salsette Coasts in their widely extending marshes. This Malacca-bean plant of the American Botanists, essentially ripal in its habitat, is an entirely distinct plant from the inland and jungle-loving marking- -nut tree. The Malacca-bean plant is known to, and described by, American Botanists as the Soft-leaved *Avicennia*. Witness the following observations of Mr. Thomas Nuttall, F.L.S. — "The *Avicennia* or Malacca-bean, according to Rheed, becomes a tall and graceful tree on the coast of India, rising to the height of 70 feet, with a trunk of 16 feet in circumsference, sustaining a pyramnidal and somewhat orbicular summit ,of dense and dark verdure. The wood is whitish, covered with a grey bark, and is employed for many economical purposes. The kernels, naturally bitter, deprived of this quality by steeping and boiling in water, are then sufficiently edible, and known to the Hindus by the name of *Caril*; an oil may also be expressed from them as from the nuts of the *Anacardium*." (Nuttall's North American Sylva, Vol. III, p. 79). In reading the above account in Mr. Nuttall's work, what puzzles me most is, why Mr. Nuttall gives *Anacardium* (*Bauhin*, Pinax p. 511; *Oepata. Rheed*, Malab. Vol. 4, p. 95, tab. 45) as one of the synonyms of *Avicennia tomentosa*. The two natural orders, namely, *Anacardiaceae* and *Verbenaceae* to which *Anacardium* and *Avicennia* respectively belong, are as far removed from each other as they possibly can be among the Flowering plants. To sum up, it is necessary to remember that the Malacca-bean plant is entirely different from the marking nut tree producing the so-called Malacca-beans of Masters.

The cotyledons of the marking-nut, when mature, yield a bland sweet oil, which is entirely free from acrid property. That the pericarp should contain a highly acrid and corrosive oil, nay, indeed, the whole plant should be more or less acrid, and that the oil of the cotyledons, on the other hand, should be perfectly harmless, is not without a parallel in the vegetable kingdom, as will be seen from my remarks on the oil obtained from the seeds of the *Moringa pterygosperma* in my last contribution to this Journal, describing that plant.

I may add one more instance of this strange botanical fact recently gathered from the writings of an American Botanist — Mr. Thomas Nuttall, F.L.S. He says that the leaves of the Western Ye tree (*Taxus occidentalis*), which is the same as the *Taxus brevifolia* of the Americans and the *Taxus baccata* of Hooker (in part Flor. Bor. Amer. 2, p. 167) "are poisonous to horned cattle and horses, though the berries are inoffensive. Cattle so affected run about in fury and delirium, and at length drop down dead. Three children, according to Dr. Percival of Manchester, were poisoned dead in a few hours by taking a small dose of the green leaves as a remedy for worms, but they appear to have suffered no pain, and after death looked as though they were in a placid sleep. The best antidotes to this poison are oily substances." (Nuttall's North American Sylva, Vol. III, p. 89).

Other instances are not wanting in the Natural Order *Anacardiaceae* itself where the cotyledons are perfectly harmless, though the other parts of the plant contain an acrid juice, as for instance the kaju-nut (cashew-nut). The oil obtained from the cotyledons of this nut is bland, and free from acridity, whereas the pericarp contains an acrid oil. Baillon mentions another plant, the *Semecarpus atra* (*Rhus atra*, Forst.) of a similar nature. The roasted seeds of this tree are eaten in New Caledonia; but the juice of the stalk, or *Nolé* resin, as it is called, is caustic and poisonous. The apple of it; besides, known as the *Nolé apple*, which is merely the fleshy peduncle, is used to prepare a fermented drink.

The most interesting part of the marking-nut tree is the pericarp of the fruit. "The pericarp contains," says J. Lapire, "32 per cent. of a vesicating oil of specific gravity 0.911, easily soluble in ether."* This volatile oil is the ingredient which gives the resinous fluid its acrid and corrosive, escharotic, or caustic property. The acrid juice

*[Journal Pharm. (3) XL., 16] — Quoted from Henry Watt's Dictionary of Chemistry, Vol. V., 1869.

of the pericarp contains *anacardic acid* and *cardol.* Besides being soluble in ether as noted by Lapire, it is soluble in alcohol and vegetable oils. *Anacardic acid* and *cardol* are readily decomposed by heat. Exposure to them produces irritant effects even from a distance in some constitutions, particularly in those which are predisposed to irritation by virtue of their tender cutaneous system. It is impossible to know beforehand which constitution might suffer and which escape. When, however, it is once found out that a particular constitution does suffer from near or from afar, it were best for such a constitution to keep aloof from the marking-nut, or its operations, when other persons are using the nut. "According to Basiner," says Dr. Lyon, "when subcutaneously injected in large doses, the yellow oily liquid known as *cardol* causes in warm-blooded animals stupor and paralysis." Medicinally given, the juice, as a whole, acts as a stimulant and narcotic. Surgeon-General Balfour notes that the juice is much administered internally to elephants. "Given in large doses," says he, "it renders these animals furious."

The volatile nature of the oil of some of the members of the *Anacardiaceæ* is very striking. Thus, for instance, referring to *Schinus molle* (false pepper), Sir Joseph Hooker, in editing Mrs. Hooker's translation of LeMaout and Decaisne's "Descriptive and Analytical Botany," adds a note that "fragments of the leaf of this plant floated on water move about by jerks owing to the discharge of a volatile oil from the tissues." Let me add that this false pepper plant is a small tree of tropical America, with a sugary edible drupe and a mastic with a slight odour of pepper (LeMaout and Decaisne, p. 363, 1873). Balfour says that the torn leaves of this plant send out the resinous matter with so great a force "as to cause a sort of spontaneous motion by the recoil." Another congener of the marking-nut known as *Melanorrhoæ usitatissima* (or *usitata* as Dr. Gregg calls it), which yields the so-called Martaban varnish, possesses a thick viscid greyish fluid, assuming a black colour on exposure. It constitutes the black varnish of the Burmese, and is extensively used by them. The point I wish to notice about this fluid black varnish is, that it has distinctly terebinthinate smell. Ordinarily the black oleo-vesinous fluid from the pericarp of the marking-nut is without any smell. "It is known as the black varnish of Sylhet" (Birdwood's Economic Products of the Bombay Presidency, p. 261). When a dried nut is held over the flame of a candle, it burns beautifully, throwing out spirts of lit volatile oil which appear in succession like bright stringed beads, dropping down copious dark thick varnish-like juice. Around the flames there is a bluish halo. As the dried fruit burns, it emits a faint smell not unlike that produced at the time of baking the cashew nut. In referring to the volatile nature of the oil issuing from other parts of the marking-nut-tree, Col. Drury and Surgeon General Balfour, in their respective works cited in this paper, make the following commong remarks — "Even the farina of the anthers of the flowers is very narcotic and irritating; people of a peculiar habit, accidentally sleeping under the tree when in blossom or even going near the flowers, are stupefied and have their faces and limbs swollen." O'Shaughnessy says that "the vapours arising from the nuts during roasting have been often known to occasion erysipelatous inflammation" (Bengal Dispensatory, p. 280). Instances of the volatile nature of the oil found in the plants of the genus *Rhus* belonging to the Natural Order *Anacardiaceæ* are to be seen in *Rhus toxicodendron* and *R. radicans*, which have an extremely dangerous milky juice. "It is capable of poisoning persons who approach such plants in hot weather" (Lindley). "It is said," notes Wight, "that merely handling the leaves of *Rhus toxicodendron* is sufficient to cause a crop of vesicles, and exposure to the volatile emanations from *R. toxicodendron* and *R. venenata* has been known to excite in susceptible persons severe indisposition, or they have had their bodies covered with a crop of pustules. The two Indian species of *Rhus*, he says, are not endowed with these acrid properties (Illustrations of Botany, Vol. I, p.183, 1840). It may be observed here that Hooker (Br. Flora, Vol. II, p. 9) says the genus *Rhus* often contains an acrid juice. He does not, however, specify which of the 12 species he describes have it. "Dr. Bigelow mentions," says Beck, "that the has known individuals badly poisoned in winter from the wood of *Rhus vernix* accidentally burnt on the fire" (Medical Jurisprudence, p. 837, 1836). It will thus be clear that it is the volatile nature of the oil which renders it dangerous even from a distance.

It will be evident from the foregoing remarks that the genus *Semecarpus* is not the only one belonging to the Natural Order *Anacardiaceæ*, which is known to affect some individuals even from a distance. Mr. Thomas Nuttall, F.L.S., an American Botanist, in his "North American Sylva" (Vol. II, pp. 121-123, 1852), mentions several other American species of the genus *Rhus* which, besides those I have mentioned, are topically poisonous. In speaking of the "Coral Sumach" (*Rhus metopium*) he makes the following remark — "Like several other native species of the genus (*Rhus*), this stately species of Sumach, a native of Jamaica, Cuba and Key West, is to some individuals poisonous to the touch. This, and the Mountain Sumach, are called in St. Domingo 'Mountain Manchiniel,' from the poisonous qualities of the juice they exude." Referring to another species, the *Rhus vernix*, to which I have just referred, he says that it "affords the Japan varnish, which oozes from incisions made in the tree, and grows thick and black when exposed to the air. It is so transparent, that when laid pure upon boxes or furniture every vein of the wood may be clearly seen. With it the Japanese varnish most of their household furniture made of wood. The milky juice of the plant stains linen a dark brown; the whole shrub like our Poison Ash (*R. venenata*), to which it is nearly allied, is in a high degree poisonous; and the poison is communicated by touching or smelling any part of it. Inflammations appear on the skin in large blotches, succeeded by pustules which rise in the inflammed parts, and filled with watery matter, attended with burning and itching, which continues for several days, after which the inflammation subsides. The extremities and glandular parts of the body are those which are most affected. Our *Rhus radicans* and *Rhus toxicodendron* (Poison Vines) operate nearly in the same way, though in a less degree than the Poison Ash or *Rhus vernix*. Many persons, however, can approach and handle these deleterious plants with impunity. One of the most dangerous species in America is the *Rhus pumila*, of Michaux, a native of North Carolina. Mr. Lyons, a well-known and assiduous collector of rare and ornamental plants, suffered extremely from its venom by merely collecting the seeds; it produced a general fever, and affected the use of his limbs for several years." Lindley observes that the *Stageneria vernicifera* of the Indian Archipelago, another member of the Anacardiaceous family which yields the black varnish called Japan lacquer, contains a resin which is extremely acrid, causing excoriations and blisters if applied to the skin. The people of Sumatra consider it dangerous even to sit or sleep beneath its shade.

It is not that every member of the Anacardiaceous family contains an acrid resinous fluid. Ondaatje, of Ceylon, for instance, says that the resin yielded by *Semecarpus Gardneri*, a native of Ceylon, is free from acridity (Pharmac. Journal. p. 819, Vol. XIII, 1882-83, 3rd Ser.).

The dirty brown soft gummy exudation from incisions made in the bark is notably tasteless and free from acrid properties; it dissolves slowly in the mouth (Brandis). This freedom of the gum from acridity is not unlike that of the pinkish gum of *Moringa pterygosperma*, nearly every part of which plant partakes of the odour of the true horse-radish or of mustard oil. It must be noted that a distinction should be made between the pure gum and the gum-resin or oleo-resin which respectively exude from different parts of the plants belonging to the Anacardiaceous family. The one is a pure gum in solid masses or tears, whereas the other is a resinous tarry oily-looking fluid, generally white in the tender parts of the *Semecarpus anacardium* plant and its young fruit.

The resinous juice of the pericarp is much used medicinally by the natives of this country. It may be named, without fear of contradiction, as the Indian *Panacea* for all sorts of complaints from the mildest to the gravest. In fact, there is no affection which the Indian flesh is heir to in which it is not, almost instinctively, used. It is often used externally with marked relief, barring individual idiosyncrasy, which renders some constitutions readily susceptible to its poisonous effects; it is not less often productive of dangerous results in other instances, the remedy proving worse than the disease itself for the cure of which it was originally used, and producing large-sloughs, far beyond expectation. Internally, the juice acts as a tonic, though at times it produces violent gastric inflammation. When administered by the mouth, the resinous juice is well dissolved in vegetable oils, milk, curds, or whey.

As an article of domestic economy the resinous juice is extremely valuable. It is commonly used by the natives for marking linen and cotton fabrics, the marks thus produced being indelible, leaving the fabric perfectly uninjured at the same time. Hence, it is that the fruit comes to be named the *Indian marking nut.* Ashes or quick-lime and water are used to fix the black colour. "The green nuts," says Roxburgh, "well pounded into a pulp, makes good lime." "The resinous juice is insoluble in water and is only diffusible in spirits of wine," adds Roxburgh, "for it soon falls to the bottom, unless the menstruum be previously alkalized; the solution then is pretty complete, and of deep colour. It sinks in, but soon unites perfectly with, expressed oils." Hence, it is that the natives use cocoanut, *teel*, or ground-nut oil to wipe off the skin recent stains of the resinous fluid, or to reduce the acridity of its poisons. Brandis says that the oil of the seeds of the marking-nut mixed with the milk of *Euphorbia* (which species, he does not mention — K.R.K.) is made into bird-lime by the wild tribes of the Satpura range in the Central Provinces.

POISONOUS PROPERTIES.

The marking-nut is a distinct vesicant of the skin, if it were nothing more. But it is even worse, as will have been gathered from the foregoing remarks. It only remains to look at its properties from the clinical point of view. Some escape its action beyond a slight irritation where the nut is locally applied. Others suffer from irritation of even untouched parts, while some others suffer violently even from a distance, though not themselves the users of the irritant fluid. Their face, their eyes, their ears become rapidly swollen, not unoften with great constitutional disturbance, followed by much prostration. The fruit is seldom, if ever, used internally for the purpose of poisoning. Jail-convicts and malingerers sometimes use it for producing blisters and ulceration of the skin, to avoid work. Dr. Norman Chevers bears me out in this assertion. "Dr. Hornigberger notes," says this experienced veteran Medico-Jurist, "that at Lahore he discovered that some of the prisoners rubbed the juice of this nut on their eyes; others had rubbed it on their bodies. The former were apparently suffering from incurable ophthalmia, the latter from a kind of ring-worm. By this means they continued to remain on the sick-list, eating and drinking without being necessitated to work." Superintendents and Medical Officers of Jails beware! Udoy Chandra Dutt also notes a similar experience as regards the jail-birds of Bengal. "It is often employed by natives," says he, "especially jail-convicts in India, to produce irritation and fictitious marks of bruises." (Materia Medica of the Hindus, p. 141, 1877). Such has been my experience in the Thana prison.

Two cases cam under my observation in the Thana Civil Hospital in 1893. In one of these cases, the deliberate application of the resinous fluid on several parts of the body produced violent symptoms; in the other case, mere pounding of the seeds on a stone was enough to cause irritation of several parts of the body. It would be useful to give briefly the notes of these cases.* They are as follow—

1ST CASE. — A female, aged 20, married, was sent to the Civil Hospital, Thana, by the Chief Constable of Salsette, on 24th April, 1893. A number of black-nut marks were found on both breasts, thighs and genitals. She stated that the marks were made by her husband by way of punishment. The parts mentioned were all swollen on the day of her admission. On the third day of the application large blisters appeared on the painted parts. Under prompt application, first of sweet oil (*teel*), then of a lotion of acetate of lead and opium, and subsequently of vaseline, the parts recovered, and the patient was discharged cured, but not before three weeks.

2ND CASE. — In this the whole face and ears were much swollen, and there were large vesicles on both the forearms, chest and abdomen. The following are brief notes of this case — The patient is a young Hindu, male, aged 24; admitted into the Civil Hospital, Thana, on 24th June, 1893. States that he had a swelling on his right forearm, for which he was recommended the external application of the marking-nut. He pounded a few nuts and used the resinous fluid on the affected part. The following day large vesicles appeared on it, as also redness and swelling on the face and ears, though they were not painted over. There was violent itching and much constitutional disturbance. The chest, abdomen, and left forearm were also swollen and red. Pus-

* The notes of these cases are recorded by my Assistants, Rajanna Lokaji and Madhavrao Shankar Sowani - K.R.K.

tules formed on these parts on the third day; the right forearm, over which the fluid was actually used, suffered most. It was considerably swollen. Within two days, under the same plan of treatment as followed in the first case, coupled with suitable constitutional treatment, the swelling of the ears, face, chest and abdomen subsided; the parts over which pustules had appeared healed up, and the man was discharged cured within ten days of the application.

The first case mentioned here is not the only one of its kind where the marking-nut is used by a cruel husband to punish a helpless wife. Dr. Lyon mentions a case (Med. Jurisprudence, p. 190,) in which a man was tried before the Bombay High Court, and "convicted of causing hurt to his wife by throwing marking-nut juice over her face, blistering of the skin and severe ophthalmia of one eye lasting for several days being the result." Dr. Wellington Gray, when acting Chemical Analyser of Bombay, came across a case where a man introduced three marking-nuts into the fenitals of his wife.

It is not uncommon to find persons painting their skins over to support false charges of assault. Dr. Newton mentions that the bruised nut applied locally to the *os uteri* is often in use amongst the native women for procuring abortion.

Dr. A. Gibson records a case in the Transactions of the Bombay Medical and Physical Society, in which a singular vesicular eruption was produced by the external application of the juice of the marking-nut (p. 117, Vol. for 1841). At page 271 of the same volume, is recorded a case in which the juice of the marking nut was used mixed with *Nereum odorum* (Marathi — *Kanher*), *Plumbago zeylanica* (Marathi - *Chitrak*), sulphate of copper, beetles (genus *Mylabris*) and a snake to boot. No wonder such a diabolical combination, as virulent in effect as it was disgusting in conception and practice, produced a fatal result!

Dr. O'Shaughnessy records his experience of the deleterious nature of the resinous juice of the marking-nut with the true spirit of a pharmacological martyr, in his Bengal Dispensatory (page 280, Ed. 1841). A minute drop of the juice placed by himself on the back of his hand occasioned "the eruption of an herpetic blotch, intensely itchy and scaling from the centre to the circumference, which did not disappear for eight months, and left a scar like that of a burn. Dr. O'Shaughnessy says that "by *boiling the nut* (the italics are mine, K.R.K.), an oil is prepared, which, when undiluted, acts as a blister." The common experience is that the process of boiling the nut is hardly necessary to bring out the blistering property. The crude juice, even as it escapes from the pericarp, rapidly produces a blister. In some constitutions the blister appears after 24 or even 48 hours, and not unfrequently, even a pustule. It is the crude juice that is generally used for medicinal or criminal purposes. The process of boiling the nut therefore, may be deemed superfluous.

Mr. C. Duran reports in the *Medical Times and Gazette* (p. 519, Vol. II, 1875), a case treated under the care of Dr. Frederick Taylor in Guy's Hospital London, for poisoning by the Indian marking-nut. The symptoms are briefly as follow — A European school-boy, aged 13, was admitted into Guy's Hospital on 18th May, 1874. He stated that he had painted the figure of an anchor on his left arm with the juice of the marking-nut *nine days* previous to admission into the hospital. The seed was given to him by a soldier returning to England from India. The juice "was not pricked in as in tattooing," but merely rubbed on. When the juice dried, it left a black stain which could not be rubbed off. During the week it caused some smarting. A full week after, the boy noticed that the painted arm was red, and that a number of small pimples had appeared. In the night his face became red and swollen, and the next day he went to the hospital. During the first night of his admission into the hospital, he rested his arm, which was covered with lint, on the abdomen and thighs. The following morning both the abdomen and thighs were red. This shows how long the acrid properties of the juice remain, and how indestructible it is even when mixed with the serum of the blood for a whole week, and how it affects adjacent parts by means of such serum even when they are protected by a thick piece of lint. The boy suffered from no constitutional disturbance. The redness of the affected parts resembled that of erysipelas. Although it was the *left* forearm that was painted over with an anchor, the lack of the *right* hand and front of the wrist "were covered with a few red spots, surmounted with minute white pustules." The face also presented a similar inflammatory condition; "the redness and swelling affected both eyes (the left being quite closed), as well as the upper lip, chin, and the back of the right cheek, where there were a few isolated, raised red spots. The colour was sharply defined, and on the chin was a yellow crust resembling the crusts of eczema. On the inner side of each thigh at the upper part was a large red, less sharply-defined patch, with one or two transparent vesicles; and on the lower part of the abdomen were a few spots resembling those on the right hand and wrist. The eruption caused considerable itching." This, and numerous other cases which have come under my observation amply illustrate the erratic nature of the poisonous volatile principle which renders the resinous juice of the marking-nut so harmful, and at times even dangerous. The boy recovered under suitable treatment and left the hospital in a fortnight. The point noticeable about this case is that a playmate of the above patient was painted with the juice at the same time, and in, consequence suffered from symptoms which Mr. J.D. Roberts, under whose care he was placed, describes as being similar to those of the cases I have cited. Mr. Duran notes that Dr. White of Harvard University characterizes the eruption as eczematous, and states that he has never seen it take on a really erysipelatous type. The duration of the disturbances generally varies from five to six weeks. (*Opus cit.*, page 520) "These cases show," observes Mr. Duran, "that the period of time which may elapse between the application of the irritant and the development of the symptoms" may be so great as to render it difficult to trace the latter and "connect them with their real cause." Dr. White is said by Mr. Duran to observe, with reference to a similar peculiarity of the American genera of the *Anacardiaceae*, that "the time required for the development of the visible manifestations of the poison upon the skin after contact with the *Rhus*,* or its emanations seems to vary greatly; for he found that, though shorter periods might occur, three, four and five days were repeatedly given, by patients especially susceptible to its action as the interval between contact and the appearance of the eruption."

In the papers published under the head of "Indian Museum at South Kensington" in the "London Pharmaceutical Journal" (p. 181,

* *Rhus toxicodendron* and *R. venenata*, already referred to in this paper.

Vol. VI, 1875-76, 3rd Series), I find the following remarks which are well worth quoting — "The marking-nut which often come to this country mixed with the myrobolans, contain between the inner and outer layers of the shell or pericarp a remarkably caustic blackish oily fluid *which is apt to blister or greatly inflame the lips of those who attempt to crack the nuts.** In persons of erysipelatous tendency, the inflammation is often considerable and even dangerous." In one of his earlier papers contributed to the "London Pharmaceutical Journal" so far back as 1878 (p. 1003, Vol. VIII, 3rd Series), Dr. Dymock observes as follows — "Garcia d'Orta remarks that the poisonous properties of the marking-nut have been much exaggerated by Serapion, and goes on to say that in Goa it is administered internally *after having been steeped in butter-milk,** and is also given as a vermifuge; and, moreover, says he, we (the Portuguese) salt the young green fruit and use them like olives." I have this to observe regarding the remarks of Garcia d'Orta. I have not the means of knowing at the present moment to what extent and where Serapion has exaggerated the poisonous properties of the marking-nut. All I can say is that those poisonous properties are very grave indeed topically in many men. If the oil of the seed is administered in Goa internally for asthma, and for the matter of that for any other complaint, or no complaint whatsoever after the seeds are steeped in butter milk, I can safely say that the oleaginous or fatty principle of the butter-milk wellnigh renders the poisonous property of the resinous juice harmless by dissolving it out, inasmuch as it is now well established that the volatile oil parts from its host (*the seed*), and leaves it harmless when much diluted in oil. Moreover, I do not believe that the Portuguese stomach is in any way different from the ordinary human stomach if it can digest without suffering the salted young green fruit of the marking-nut as a substitute for olives, with the prior aid of common salt, which possibly renders the acrid properties of the seed inert, by dissolving out the poisonous element and changing it chemically. During the process of salting probably the volatile principle escapes. There is no reason why the chlorine of the chloride of sodium should not act upon the constituent elements of the marking-nut.

* The italics are mine - K.R.K.

In my humble opinion the "Pharmacopoeia of India" in no way exaggerates the deleterious properties of the marking-nut. It is safer to emphasize them in the present state of our knowledge as derived from books and from practical experience. Dr. Dymock distinctly says that when the nut is given internally "the juice of the pericarp is mixed with oil or melted butter." This is his last utterance on the subject ("Pharmacographia Indica," Vol. I, page 390. "In the Concan," says he, "a single fruit is heated in the flame of a lamp and the oil allowed to drop in a quarter seer of milk." This drawing of the bare black juice, deprived of its obnoxious, volatile, poisonous element, removes a great deal of the sting of the poison if not the whole of it. The poisonous element lies in the volatile oil, which, like all other volatile substances, is dispersed by means of actual heat, or when exposed to air.

ADDENDUM. — While the first proofs of the above contribution were being examined by me, with the assistance of Mr. Mahadeo Shankar Sowani, my Civil Hospital Assistant, he told me that Dr. Ganesh Krishna Garde, L.M. & S., of Poona, could supply me with some important notes of cases of poisoning by marking-nut which he had met with in his practice. I have known Dr. Garde by reputation and by personal acquaintance as a medical man of great research and possessing extensive knowledge of Indian drugs. He is well read in ancient Hindu medicine. His experience therefore is worth recording. On application he has promptly supplied me with the following information which I produce *in extenso* in his own words—

"Marking-nut has a special affinity for the skin, and its action on it must be looked upon as specific, as it invariably affects that tissue whether applied externally, taken internally or in whatever way absorbed into the blood. I quote below one or two of several cases that have come under my observation.

CASE No. 1. - A Parsee gentleman came to me one morning complaining of big swellings on the face, forearms, the back of the wrists and chest. There was no tingling or itching, but a slight dull tensive pain on tender parts like cheeks and alae of the nose, &c. Some of the swellings had a reddish discoloration but others were of the natural hue. My former experience at once suggested to me the idea that these swellings might be due to marking-nut in some way absorbed into the blood; and so I questioned the patient whether he had ever come in contact with marking-nut in some way or other. After repeated questionings the patient recollected that he was lately using for some time a tooth-powder which contained a charcoal of these nuts. On learning this history I told him to stop using that tooth-powder and gave him one of my usual antidotes for application to the swellings which gradually disappeared in a few days. *Lininentum calcis* or a weak ointment of Bicarbonate of Soda are generally serviceable when these swellings are fresh, but when they get older I generally prefer an ointment of *Rhus toxicodendron* on Homoeopathic principles, for *Rhus toxicodendron* has a similar action on the skin and a weak ointment of its tincture very rapidly disperses all anacardium irritation and swellings and ulcerations (caused by local application of the nut.). I have lately learnt that the ashes of the bark of the marking nut tree is a good application for all the cutaneous lesions caused by the nut.

"CASE No. 2. — A few months ago a man came to me from Kolhapur with a hard, reddish and slightly painful swelling on the tip and alae of the nose, a part of the cheeks and upper lip. It had remained persistently, shifting its place a little now and then, and becoming aggravated or ameliorated by various external applications prescribed by Doctors and Vaidyas but never disappearing altogether. As it had remained nearly two years so obstinately he was alarmed and came here for consultation. He told me that the hard, reddish swelling on the nose and surrounding parts was diagnosed as lupus by a renowned medical man but his treatment had failed. In investigating the causes of the swelling I happened to ask him, among other questions, whether he had ever used marking-nut internally. To this he answered affirmatively, and said that he had on several occasions consumed a nostrum containing these nuts prescribed for dyspepsia, constipation and general debility. This at once put me on the right track and my diagnosis of the case was Anacardium rash. But I regret to say that the diagnosis of the case was not completed by treatment as the patient did not remain here any longer.

"I have had several opportunities of observing the effects of marking-nuts and my long observation leads me to believe that *Anacardium orientale* is a specific dermato-irritant, its special affinity being for the skin of the face, alae nasi, lobes of the ear, and lips and to some extent the trunk and upper extremities. Of course locally applied it may affect any part of the skin. The swellings over the nose, &c., have a striking resemblance to leprosy, and this may account for its recommendation for that disease in native medicine. Indeed modern Homæopathic therapeutics recommeend it against leprosy on account of its Homæopathicity to that disease. It is said also that long continued internal use of the drug causes failure of memory and some forms of unsoundness of mind."

In a further note Dr. Garde adds the following remarks — "Marking-nut is one of those drugs like Quinine, Nux-Vomica, Ipecacuanha for which certain individuals show a great idiosyncrasy. I have seen people eating these nuts with cocoanut kernel without any harm, and many have used a decoction of about a dozen nuts quite safely. There are, on the other hand, some people, especially children who get their face swollen simply by being exposed to the effluvia of the nut, evolved when it is being broken for external application. These effluvia have been known to affect this way even from a considerable distance. The people of Gujrat and Kathyawar are more sensitive to its action than the Dekkanees. Darker races seem to resist its action; and people of fair complexion are very sensitive to its action."

I am all the more gratefull to Dr. Garde for these interesting remarks, especially as he is convalescent from a serious illeness, and has promptly complied with my request in a very trying condition. Although I am not prepared to accept his Homoeopathic views, I have let him speak in his own way *in extenso*. There is one remark I would make, and that is with reference to his observation that the long continued internal use of the nut causes failure of memory, and brings on some forms of unsoundness of mind. In the body of the paper I have already observed that I have seen it used in a case of leprosy for improving lost memory. I have also quoted from Masters the view that is stimulates the mental powers, and especially the memory.

Reprinted from
Journal Bombay Natural History Society
Vol. 9(4): pp. 345-365. 1894

PART XII.

PLUMBAGO ROSEA- (*Linn.*).
Natural Order -PLUMBAGINEAE.
MARATHI- लाल चित्रक LAL CHITRAK

An envergreen perennial shrub, 2-4 feet high, "very rarely annual," (Boissier). The plant when grown in gardens and in conservatories in much more handsome than when it grows under a blazing sun, whereby it s leaves invariably suffer in form and substance.

STEM. -Suffruticose, procumbent near the root, sometimes scandent, jointed where the leaves are inserted, woody below herbaceous above; thin, long, roundish, longitudinally striated; near the joints the green colour of the stem is darker green, lined with red; usually not much branched. Sometimes instead of a single stem a number of branches shoot out direct from the tortous roots. Some of these branches are erect, others procumbent, trailing along the ground.

Usually however, especially in new plants, there is a single stem from which several branches arise. Young stems and branches contain pith in the centre.

BRANCHES. -Bifarious, diffuse; striated longitudinally; generally borne on the upper part of the stem, less numerous on the lower. Length of branches sometimes three to four feet at a stretch. The stem when in good condition bears leaves 1-1/2-2 1/4 feet asunder. Each articulation bears a solitary leaf. The procumbent branches, when they reach the ground, bear fresh rootlets from the joints.

ROOT.-Fibrous, outwardly dusky yellow, or greenish according to Rumphius. On section it looks yellowish with a brown tinge in the line of the axis as seen in figure 5, of Plante N, accompanying this letter-press. The cross-section depicts the appearance of the root in its fresh condition; on drying it shrivels up. When mature it is woody and solid. Some say it has a disagreeable odour. As compared with the general size of the plant, the root is thick and strong; nearer the stem it is nodose and sinuous sending down into the ground many thick fibrillae. Sometimes the root is two feet long; sometimes it sinks much deeper ito the soil, so much so that it is difficult to extirpate it.

LEAVES.- Entire, alternate; slightly drooping; glabrous; undulate or wavy; elliptic, ovate, lanceolate, or oblong-rotund, as Rhhede calls them; striated by means of prominent ribs on the ventral suface; midrib markedly reddish; the principal nerves are similarly reddish or even faintly so on the ventral surface in tender leaves, they are greenish on the under or dorsal surface. The margin of tender leaves is distinctly tinged red all round. The leaves vary in colour from dark green to light green. The tender leaves are generally light green; the older dark green. Leaves growing in shady places are of ligher green.

PETIOLES. - Very short; channelled; sometimes amplexicaul, sometimes absent; marginal tinged red. When the petiole is absent the leaves are sometimes amplexicaul or semi-amplexicaul.

The apex of young leaves is delicately mucronate. To see this, one has to examine the leaf buds as they expand at the growing point. The older leaves are obtuse at the apex. The under or dorsal surface of the leaves is generally of a lighter colour than the upper or ventral.

VERNATION.- Involute.

STIPULES. - Absent.

FLOWERS.- Usually about 1 1/2-2 inches in length; not persistent; falling before the fruit is mature or even formed; situated on spikes.

INFLORESCENCE. - In terminal or axillary spikes (racemes). When the spikes are axillary, they arise from the upper axils of branches. Spikes usually simple, 1-2 feet long. Rachis of the spike glabrous. The flowers are arranged on the rachis alternately an inch or more apart, sometimes only on one side of the rachis,* thus forming unilaterla spikes or racemes. The flowers are never crowded together. They are sometimes erect, sometimes transverse, and sometimes bent downward or slightly drooping.

ÆSTIVATION. - Of Calyx, plicate (Blume); of petals contorted; often sinistrorse, sometimes dextrorse.

The flowering time is said by Blume to be in May and June. On this side of India and in European conservatories the plant is well known to flower more than once a year. There is, however, a short resting period after the whole plant has once flowered. The blossom does not all appear at once on the whole plant, but branch after branch throws out spikes which bear flowers in due time. Rheede says that the plant flowers throughout the year, but especially in the rainy season.

BRACTS AND BRACTEOLES. - Shorter than the calyx.

BRACTS. - 2, superior lateral; 3 says Roxburgh. Often scarious on the margin; ovate or ovate-oblong.

BRACTEOLES. - 2, "conform with the bracts" (Kurz). Rheede describes 3 small bracteoles, glabrous foliaceous. Rheede's description seems to me to be more accurate; at any rate 3 bracteoles are more common. The bracteoles are persistent; even after the flowers have fallen, for a long time the bracteoles are to be seen intact, of reddish-brown colour. They are greenish when young. Each set of bracteoles, whether 2 or 3, contains a single flower. The flowers are without any odour.

* The flowers, says Sir William Hooker, "are remarkably secund, that is, turned to one side" (Curtis's Bot. Magazine, Tab. 5363). This is well illustrated in the plate given under that number in Curtis. Curtis's plate 230 in an earlier volume of his Bot. Magazine bears bilateral flowers like those in plate N of my series.

CALYX.- Inferior; gamosepalous, tubular; herbaceous, covered with stalked sticky scarlet glands throughout its length; 5 -10 ribbed; often hyaline between the ribs. Mouth of the tube funnel- shaped, scarious, very slightly 5-dentate. Limb not persistent; but more so than the corolla. Upper half bright red, often deep rose-coloured; lower half brown-red, with just a shade of green near the insertion of the bracteoles. The calyx is distinctly non-accrescent, nearly always sessile. The pedical if present is very short, oblong, and deeply imbedded in the bracteoles. The calyx of young buds is brightx green; the stipitate green on it are also green. The viscid muscus secreted by the stalked gland, gladular hairs as some call them, does not dry- (Boisser). Recent rearches on these glands will be refferred to under the heading of *Remarks* later on.

COROLLA.- Bright red, oftem deep rose-colored; hypogynous; hypocra- teriform; tube long, slender, slightly angulate, generally lighter in colore than the lobes at its lower third. Lobes 5, patent, sometimes equal, sometimes slightly unequal. The form of the individual lobes varies much even in the same flower, being sometimes round, sometimes obovate, but usually obcordate. The midrib of the petals is distinctly paler and prominent on the under surface, ending sometimes in a sharp point. Hence Boisser calls the petals "unguiculate;" sometimes the petal is blunt at the apex, hence assuming the obovate form; sometimes it is distinctly notched, thus giving the petal an obcordate appearance. The petals are sometimes faintly striate. When fully open the corollat resenblese that of the Jasmine. The circumference of the expanded petals taken together is an inch or even more. The corolla falls comparatively earlier than the calyx.

STAMENS.- 5, opposite the petals; free; very fine and delicate; pale rose-coloured, partly whitish; slightly rising above the "throat of the corolla-tube, and looking purplish.

FILAMENTS. - Filiform; linear; dilated at base and inserted at the base of the petals "into the tops of the 5 nectarial scales"- (Rox-burgh).

ANTHERS. - Oblong; introrse; anther-cells separated at the base, and opening by longitudinal slits.

POLLEN-GRAINS. - Ovte.

HYPOGYNOUS DISK. -Absent.

PISTIL-

OVARY.- Free; sessile; superior; 1-celled; 5-angular above, or "5-gibbous" as Boisser terms it; narrowed at apex.

STYLE.- Single, slender, whitish; with 5 small segments, quite indiscernible and alternating with the stamens. Base of the style hairy. The union of the segments is a diagnostic character of this plant.

STIGMAS.- 5, "capillary, furnished on their inner surface with several lines of glands" (La M. and Dec., Mrs. Hooker's Translation, page 525).

OVULE. - 1 pendulous from a basal ascending funicle; anatropal.

FRUIT.- A membranous capsule, circumciss or rupturingx near the apex;' included in the tubular persistent calyx. Blume says that the dehiscence is valvate; others say that the pericarp of the fruit dehisces in a subvalvate manner.

SEED.- Cylinderical or oblong-rotund, with 5 longitudinal striae; pendulous; "inverted sometimes, appearing erect by the union of the funicle with its integuments." - (La M. and Dec.) The seed is green when young, black when mature, adnate to the integument at its upper extremity.

INTEGUMENT. - Simple.

ALBUMEN. - Farinaceous; slight; separated into two parts by the embryo.

EMBRYO. - Within the albumen; straight.

RADICLE. - Superior; Short.

COTYLEDONS. - Flat.

PLUMULE. -Slightly rough.

REMARKS

The plant has the following synonyms :

1. *Thela coccinea* (Loureiro, Flora Cochinchinensis, vol. I, p. 119, 1790).

2. *Thalia coccinea* (Loureiro, Flor. Cochin., Ed. Willd , vol. I, p. 147).- Curtis Bot. Magazine, Tab. 5363, 1st February, 1863.

3. *Plumbago coccinea* (Boisser in D.C. Prodr., vol. 12, p. 693).

4. *Plumbagidium roseum* (Spach, Veg. Phan., Vol. 10, p. 339).

5. "*Radix vesicatoria*" (Rumphius Herb. Amboiensis, lib. IX, cap. 68, p. 453).

In writing the above description of this plant, I have mainly followed Rheede, Rumphius and Boisser, who show a very accurate Knowledge of it. There are many points, however, in their remarks, as in those of other writers to be named hereafter, which require special notice; and I shall refer to them under this head as occasion requires. First then, with regard to its name, I have adhered to that given by Linnaeus. Boisser observes that the specific name *Thela coccinea** had to be substituted by Loureiro for *Plumbago rosea* of Linnaeus (rosea meaning rose-coloured, pink or rosy), because the colour of the species we are considering was deeper than that of the pinkest coloured rose. It was distinctly scarlet. For, says Boisser, "the corolla is never rosy in this plant."† To this, however, Sir William Hooker, presumbly the Writer of the letter- press accompanying the excellent plate in Curtis's Botanical magazine under No. 5363, gives a very pertinent reply , albeit ingenious, in the following words :— "The flowers are one and a quarter inch long in the tube and more than an inch broad, remakably secund, and the colour is bright brick-red, partaking of nothing of that purplish hue which induced Linnaeus to call the species '*Rosea*'. Perhaps Loureiro and Boisser has this variety (or this coloured variety at least) in view when they gave it the name of *Coccinea*, but the difference is hardly such as to justify the change of the old Linnaean and well- established name of *Plumbago rosea* to *P. coccinea* as Boisser has done. Popularly, too, the rose is red."

This reminds me of my school-days, when every class-book I bought I scribbled with the couplets :-

"The grass is green,
The rose is red,
This book is mine
Tili I am dead;
Steal it not for fear of shame,
For under lies the owner's name."

Of the authorship of these lines I am still in the dark as muchas I am with regard to their pretension to verse. The conclusion I draw,

* Thela, from the Greek Thelé = papilla, so named from its papillose calyx (J.D. Loureiro); Coccinea meaning red or scarlet.

† To quote Boissier's own words – "Corolla numquam in hâc plantâ rosea est." D.C. Prodr., vol. xii, p. 693.

however, is the same as that drawn by Sir William Hooker, namely, that what is rosy may be considered red. Examining furhter ther aptness of the synonyms mentioned above, I find in a recent edition of Johnson's Gardener's Dictionary (by Wright and Dewar, 1894, pp. 771-772) that, besides the species mentioned as *Plumbaago rosea*, bearing red flowers, another distinct species is given under the name of *Plumbago rosea coccinea*, which has larger and brighter flowers than *P. rosea*, and which is said to be a native of the Neilgherries. This is evidently done on the authority of Sir William Hooker's letter press accompanying plate 5363 again referred to later on. In Paxton's Botanical Dictionary (Hereman's edition, 1868), opn the other hand, I find that *Plumbago rosea* and *Plumbago coccinea* are mentioned as two distinct species. The latter is said to be scarlet-flowered and a native of the Neilgherries, whereas the former is red-flowered and found throught the East Indies. On referring to the Species Plantarum of Linnaeus (Willdenow's Edition, Berlin, 1797 , vol. 1, p. 838), I find that the species described by Burmann as *Plumbago zeylanica* is synonymous with *Plumbago rosea*. This *Plumbago zeylanica* of Burmann is a distinctly red- flowered Plant. In the *"Flora Zeylanica,"* which Linnaeus published in 1747, the same statement is made (*vide* p. 30). I mention this here particularly, because I wish that the red-flowered *Plumbago zeylanica* of Burmann should not be confounded with what we now-a-days know as *Plumbago zeylanica*, Linn., which is a white-flowered variety. I presume that this white flowered species is the *Plumbago zeylanesis* of Burmann. Asa Gray prefers to call the red-flowered plumbago of America *P. coccinea*.*

The plant we are considering is easy of propagation by means of cuttings. The cuttings root very readily in the rains. The evide ntly require much moisture and grow in almost any soil; they prefer, however, a sandy soil, and thrive, both as regards leaves and flowers, in cool and shady places. For I have observed that plants growing in the open suffer very much in foliage and blossom and never fruit on this side of India. Firminger observes that "the beauty of the plant is much impaired by many of the leaves appearing generally in a decayed condition; requires shade and moisture." (Manual of Gardening, 1890, p. 543).

* Field, Forest and Garden Botany of the United States, 1870, p. 222.

Next with regard to the habitat of the plant. So far as I know, it is not found in Bombay growing wild. It is essentially a garden variety on this side of India. The *Plumbagineae* as a class are said by Mr. Clarke to grow mostly in maritime and saline places, and the genus *Plumbago.* is distributed throughout the warm regions of both the hemispheres. But it is questionable whether *Plumbago rosea* grows wild extensively, if anywhere at all, in Western India. Kurz thinks it is not wild in Burma; Thwaites thinks it is not wild in Ceylon; so says Clarke. Loureiro says that both the species of *Plumbago, i.e.,* the red and the white, grow among the hedges of Indian bamboos, over which they the plant known as *Plumbago auriculata* (probably from its amplexicaul petiole), which is the same as *Plumbago zeylanica* of Linnaeus, very largely grows in the marshes of Batavia (Bijdragen Flor. Nederland Indie, vol. III, p. 735, Buitenzorg). Kurz says that "*Plumbago rosea* is not unfrequent in the moister mixed forests of Pegu, Yomah, and the martaban Hills down to Tenasserim; also Ava." He also adds that the plant is oftem cultivated and springs up "in toungyas, * along the river banks, & c." Rheede has noted that the plant grows in sandy places. In Bentham and Mueller's Flora Australiensis (vol. IV, p. 267), I find the following remark regarding P. rosea :- "Another Indian species, distinguished chiefly by its red flowers and more herbaceous stem, *P. rosea,* Linn., is said to have establihsed itself near Rockhampton, in queensland, as an escape from gardens." The question is whether this happens by means of cuttings or by the dissemination or actual planting of the seed. If the plant seeds in Australia at all, it is a remarkable fact- for in Bombay I have never known it to seed. There is this thing to be observed - that it is unquestionable it seeds on the Neilgherry Hills; for witness what is noted in Curtis' Botanical Magazine, plate 5363, later-press. It is evident from this plate, depicting a very handsome and substantial blossom, that Sir William Hooker considers the plant depicted theron is a variety, for he calls it "*variety coccinea*," of *Plumbago rosea*, which Johnson refers in his dictionary, as quoted in the foregoing remarks: "The plant," says Sir

**Toung-ya* (Burmese) - a form of hill-cultivation carried on by burning the jungles- (Balfour)

William, "has long been in cultivation at Kew, and Messrs. Veitch have sent us *specimens raised from seeds received from the Neilgherries.*" The plant is such an exceptionally good illustration of stove-culture that few will fail to recognize the value and extreme beauty of *Plumbago rosea* as a garden ornament. Under the burning sun of India it can scarcely hope to bear leaves "measuring six inches in length and three in breadth.." or panicles "two or more feet long" as depicted in plate 5363. the plant requires ample shade and moisture for healthy growth. Sir William's illustration is a vivid example of what the flower and foliage can be where the plant is not called upon to cope with the horrible heat and damaging dryness of the Indian Sky.

It is difficult to understand the remark made by Mr. C.B. Clarke, the writer of the article *Plumbagineae* in vol. III of Hooker's "Flora of British India" (vide p. 481) with reference to *P. rosea.* It runs as follows :- "Altogether resembling *P. zeylanica* and perhaps only a cultivated variety of it." To all appearance the plant I am describing is essentially a distinct species, in which the red colour is markedly prevalent, not only in the flowers and floral envelopes and their appendages, but even in the midrib and margin of the leaves and also in the leaf-appendages. The following conspectus of the two species given by Kurz* will show the differential characteristics of the two species:-

Spikes glandular, pubescent; corolla white, bract ovate, leafy; bractlets subulate . . . . *Plumbago zeylanica.*

Spikes glabrous or nearly so; corolla rose-coloured or scarlet; bract ovate, oblong, scarious, brown; bractlets conform with the bracts . . . . *Plumbago rosea.*

Roxburgh has pointedly remarked that the specific distinguishing mark between *P. zeylanica* and *P. rosea,* according to his observation, depends on the racemes and bracts, *colour not being a specific mark.*†

"The bracts of *P. rosea,*" says Roxburgh, "are three-fold, one-flowered, equal, smooth; the inner ort upper two united between the flower and rachis." The bracts of *P.zeylanica,* on the other hand, "are three fold, one flowered; the outer one is ten times larger than the lateral one; they are covered with the same glutem as the peduncle of the raceme; sometimes there is a fourth linear bract pressing the calyx."

* Contribution towards a knowledge of the Burmese Flora, pp. 217, 218, Journals As. Soc., Bengal, No. II, 1877.

† Roxburgh's Flora Indica, p. 155, Calcutta Edition, 1874.

With such distinctive specific differences which cannot be considered to be mere changes due to cultivation, it is somewhat puzzling to meet with the remark of Mr. Clarke in our standard work on Indian Botany.

In describing the root of *Plumbago rosea*, so far back as 1876,* Dr. Dymock says that "the root is reddish-brown externally, *transversely fissured, thick* and *Suberous; internally it is white when first removed, but soon turns red upon exposure to air.*" (The italics are mine- K.R.K.) It wll be seen from my transverse section of the root (Fig. 5, Plate N) t hat the color is yellow. It is only the central part that is brownish -red. This central portion consists of "a woody column," to use Dr. Dymock's own words,"with a very large vascular system surrounded by a bark almost entirely made up of thin-walled parenchymatous cells loaded with large oil-globules, and containing some starch." Jaikisson Indraji notes that "the roots when fresh exude a yellow juice." This would account for the yellow colour of the section of the fresh root depicted in my Plate N. Contrast with this the statement of Dr. Dymock† regarding *P. europaea* that Prof. Fluckiger and Mr. Gerock found its root when freshly cut to be nearly devoid of colour, but on exposure it immediately assumed a yellow hue. I may add from personal observation that the root of *Plumbago rosea* is naturally yellow on section, even when taken fresh from the ground.

With regard to the habit of the plant, Asa Gray's *Plumbago coccinea* is a climber. The Indian variety is not always so, but can be trained as a climber if there is suitable trelliswork to support it.

The cotyledons, according to Gaertner, are foliaceous, and ovate-oblong; the embryo is slightly yellow; the albumen perfectly while; and the integument of the seed membranceous. (De Fruct. et Semin., vol. I, p. 234). Judging from the dehiscence of the fruit figured on Tab. 50 (Gaertner), it apperas that the valves of the capsule begin separating at the apex.

With regard to the nature of the glands found on *Plumbago rosea*. I cannot do better than refer to the elaborato researches of Dr. John Wilson, Lecturer on Botany in the University of St. Andrews, published in the Annals of Botany‡. As Dr. Wilson rightly observes, Boissier

* Pharmaceutical Journal and Transactions, vol. vii, 3rd series, 1876-77, pages 190

† Pharmacographia Indica, vol. ii, p. 333.

‡ Vol. iv, No. xiv, pp. 231-255, May, 1890

does not touch the question of glands in his exhaustive monographs on the *Plumbago* in De Candolle's Prodromus, freely utilized by me in compiling the foregoing remarks. No mention is made of the glands in Bentham and Hooker's *Genera Plantarum.* Nor is there in any author among the many that I have consulted any special mention of the structure and varieties of these glands. The contribution of Dr. Wilson, therefore, is a distinct advance upon the previous knowledge of the morphology of the *Plumbaginea*; he therefore deserves the gratitude of every special student of this order for this elaborate researches. "The study of the vegetative organs of any member of the family" (*Plumbaginear*), says he, "is not proceeded far when one finds studded over the leaves, stem , &c., sunk oval or circular glands with characteristically cruciform diametrical partitions." They possess, what Dr. Wilson calls the "chalk- secreting" function, to which are due the mineral substances known as "chalk-scales" secreted by certain species of the *Plumbaginear.* Mettenius having described and figured these glands, Dr. Wilson calls them, for the sake of brevity, *Mettenian glands* (Glands of Mettenius), and as such they will be referred to in the following observations. The structure of these glands may be given as follows, as described by Maury and quoted in the original French quotation cited by Dr. Wilson.*

"The *Primitive or mother cell* is divided mainly into-four simple cells each of which latter is a secretory cell. Their product (or secretion) is stored up in the intercellular space and is thrown out (or excreted) in consequence of the tension of the cells lying close to the lower part."

These glands, observes Dr. Wilson, in general "occur in longitudinal depressions on the stems of Plumbago s giving rise to a striated appearance when calcium carbonate is secreted." They are also present "on the leaves ad bracts of all and sepals of many of the species of *** *Plumbago* *** examined." They also occur on the cotyledons of the genus *Plumbago, besides those of several other general of the Plumbaginea,* "even before the light is reached by the germ inating seedlings" (Wilson, p.240op.cit.) I have often observed a

* Translated by me with the assistance of Mr. J. Mittra, Barrister-at Law, Nagpore

whitish crust on some of the leaves of *Plumbago rosea*. But I have never been able to account for its appearance there. Viewing the crust now in the light of Dr. Wilson's researches, I am able to attribute the occurrence of it to *Mittenian* glands. I may observe here that the critical eye of Mr. Jaikisson, Indraji, Of Porebunder, has not failed to notice this crust. In a Gujarati work, entitled Arya Aushadh, published in 1889 by Dr. Virji Jhina Raval, L.M. & S., in describing Chitrak, it is stated that in the space between the amplexicaul petiole and the stem a kind of saline substance is found which, he says, is used medicinally (p. 243). Mr. Jaikisson Indraji, in describing the leaves of *Plumbago rosea*, rays that "they are covered with a white or brownish bloom." When I read this statement eight years ago, and observed later on for myself the white crust or bloom as he calls it, I was disposed to attribute it to external agencies. I thought that the saline deposit was due to salts contained in the water sprinkled over the plant in the process of watering. But, thanks to the researches of Dr. Wilson , this calcarious deposit has now a morphological interest, readily affording they happy solution of a question which has struck very few before, but which now appears to be a special phenomenon connected with the glandular appendages of the genus *Plumbago*. Were it not for the explanation given by Dr. Wilson, I should still have gone on imagining that the white crust spots seen on many leaves in *Plumbago rosea* were merely due to the watering of the plants and to the deposit of the salines from the watering fluid rather than to the inherent power possessed by the glands of secreting saline ingredients by virtue of the chemical process going on in their ultimate cells. Dr. Wilson has now shown us that the so called *Mittenian* glands which the Plumbagos contain in large numbers are the producers of the white saline crust, which consists of calcium carbonate.

The next set of glands seen in the Natural Order *Plumbaginear* constitutes, what Dr. Wilson calls, "mucilage-glands." A few of them are seen in the leaf-axils of *Plumbago rosea*, the largest being 0.38 mm. in diameter. The stalked glands on the exterior of the calyx of *Plumbago rosea* belong to this class. They are well known for the viscidity of their mucilaginous secretion. They are "spherical, bright red, and formed of very numerous polygonal cells. When fully developed they are very opaque, and their whole structure remarkably tough. The stalk penetrates a considerable distance into the fully formed head, so as to make the base of the gland dome-shaped or conical. Younger and smaller external glands are fairly translucent. The stalks of almost sessile glands are composed of a few comparatively large cells." Sessile glands are also to be seen in *Plumbago rosea* on the inner face of the sepals. Their basal cells are two. "Their hormology with the stalked glands on the exterior of the calyx," Says Dr. Wilson, "traced through the smaller and almost sessile ones there, cannot be doubted. Further, the sessile glands do not offer any point of material distinction from normal mucilage glands of the leaves."

POISONOUS PROPERTIES.

The Plant which Rumphius has described so graphically and fully under the name of *Radix uesicatorius** (the blistering root), is unmistakably the *Plumbago rosea* of Linnaeus. That this name has not been given without reason will be apparent from the following observations. It is the root of the plant which is the most obnoxious of its entire organization. Its juice is very acridly pungent t o taste, and produces in the tongue a severe lancinating pain. To do this it must be fresh. The juice distinctly blisters the skin and otherwise irritates it as violently and briskly as cantharides. Within six to eight hours, says Rumphius, the skin becomes red and covered with bleds. Rheede observes that the leaves and stems are also pungent to taste, but are not so virulent as the root. More than one writer has said that the virulent effects are more marked when the juice is taken from a fresh root. Why is it so? Is it that the irritant property is due to an evanescent active principle, an essential oil or ether which deteriorates in course of time as the root dries up? Seventy years ago M. Dulong isolated an acrid active principle called *Plumbagin* from *Plumbago europaea*, the only European congener of *Plumbago rosea*. From the researches of O'Shaughnessy it appears that the same active principle is found in *Plumbago rosea*. It abounds chiefly in the bark of the root. "It occurs in brilliant yellow crystals, of

* Herb. Amb., Lib. ix, cap. 68, p. 452

† Hortus Malabaricus, vol. x, t. 9, p. 17

sweetish, but acrid and hot taste, easily fusible, partially volatile, scantily soluble in cold, freely in boiling water, very soluble in alcohol and ether, in the concentrated acids, and in alkaline solutions, with which it strikes a deep red colour; or with hydrate of alumina* or subacetate of lead a crimson red coloured precipitate is formed, which constitutes a very valuable and delicate test, available in many medico-legal inquiries."† In the light of these researches of O'Shaughnessy it is important to remember that happily for toxicologists, *Plumbagin* is one of the few vegetable poisons which can be detected by chemical analysis as a matter of certainty. "Two grains of the powdered bark." says Dr. O'Shaughnessy, "may be detected in a pint of a mixture of milk, blood and various articles of food."

The careful researches of Dr. O'Shaughenessy are further emphasized by Dr. Dymock and his conferest in describing the chemical composition of *Plumbago rosea.* They say that "the activity of the drug depends upon the presence of *Plumbagin.*" It crystallizes, they add, "in delicate needles or prisms, often grouped in tufts; has a styptic saccharine taste; melts very easily, and partly volatalizes unaltered when heated. It is neutral, nearly insoluble and ether. It dissolves with yellow colour in strong sulphuric acid, and fuming nitric acid, and it precipitated by water in yellow flocks. Alkalies change the colour of the solution to a fine cherry-red; acids restore the yellow colour." The extreme importance of these tests, from a toxicological point of view, is my only apology for thus extensively reproducing the remarks contained in the "Pharmacographia Indica" of Dr. Dymock and his collegues. Dr. Lyon§ makes a few remarks which may well be reproduced here:- "Plumbagin treated with cautic potash solution dissolves, forming a bright crimson liquid. Hydrochloric acid added to this changes the colour to yellow, and on standing the liquid deposits yellow flocculi of Plumbagin, which may be separated by shaking the

* According to modern chemical nomenclature Aluminium trihydrate and tribasic acetate of lead respectively.

† O'Shaughnessy's Bengal Dispensatory, p. 509, 1841.

‡ Pharmacographia Indica, vol. ii, p. 332, 1891

§ Medical Jurisprudence for India, p. 219, 1889.

acidified fluid with ether. An alcoholic solution of Plumbagin gives a crimson precipitate with solution of basic acetate of lead." Dr. Lyon further adds an important note which is useful for analytical purposes :- "A colouring matter obtained from rhubarb root gives, with caustic potash solution, a crimson colour similar to that given by Plumbagin. An alcoholic solution of it, however, gives, with acetate of lead solution, an orange precipitate, not a crimson precipitate like Plumbagin."

There is yet another work, and a more recent one, I should like to refer to in connection with the active principle hitherto known as *Plumbagin*. It is the Dictionary of the active Principles of Plants, compiled and published by Charles E. Sohn, F.I.C., F.C.S., last year. At page 10 of this valuable work Dulong's *Plumbagin* is otherwise named *Ophioxylin* ($C_{16} H_{13} O_6$, or $C_{48} H_{39} O_{18}$). It is said by Sohn to resemble *Juglone*.* It is one of the sctive principles, found in some of the representatives of the Natural Order *Apocynaceae*, as, for instance, *Rauwolfia* (Ophioxylon) *Serpentina* and *trifoliata*. It is seen in the form of orange crystals of the tetragonal system, the melting point of which is 71.8. No wonder, then, that in India the active principle contained in those roots is of an evanescent nature where the normal temperature is, as a rule, above the melting point above indicated, except in the cold weather. This leads me to consider here the following remark made in the "Pharmacographia Indica" (pp. 332-333) by Dr. Dymock and his collegues, namely — "M. Greshoff, who has been investigating the chemistry of the medicinal plants of Java***, is of opinion that the roots (supposed to be from *Rauwolfia serpentina*) examined by Wefers Bettink *** were really those of *Plumbago rosea*. Professor Bettink extracted with chloroform a yellow crystalline principle, apparently the Plumbagin of Dulong. Had M. Greshoff the benefit of Sohn's recent researches, I am positive he would have come the conclusion that they both contain the same active principle;

* *Note* : – "Vogel and Reischauer discovered a substance which crystallized in reddish yellow needles in the green outer shell of the unripe walnut (*Juglans regia*), and gave to it the name of *nucine*, and afterwards changed to *juglane*. C. Reischauer then pointed out that this substance was probably identical with the *regianine* which had ben prepared from the mutshells by Phipson, and this was confirmed by the latter chemist." - Treatise on Chemistry, Roscoe and Schorlemmer, vol. iii, Organic Chem., part vi, p. 137.

and perhaps would not have said that, because the roots of *Rauwolfia serpentina* contained the active principle known as Dulong's Plumbagin, they were the roots of *Plumbago rosea*. As a matter of ract, from the statement of Sohn we know now, in 1895, that they both contain Dulong's Plumbagin which Sohn calls Ophioxylin.

So far, for the active principle of *Plumbago rosea*. Let me now go back to the researches of O'Shaughnessy, the founder of our accurate toxicological knowledge regarding the plant we are considering. Although he avowedly speaks from a clinical experience of "from 300 to 400 cases," Dr. O'Shaughnessy is modest enough to say in the very next sentence as follows:" - "Our experiments lead us to believe that the blistering quality resides in the Plumbagin, but on this we are as yet unable to speak in positive terms." This is truly characteristic of the humility of a thoroughly scientific mind. Dr. O'Shaughnessy was fully aware of the influence his conclusions would exerciese for good or for evil on the readers of his words for countless generations yet to be. Had he been aware of what the ancient Hindu writers had said about the plant hundreds of years before him, I am sure he would have been more positive in his assertion regarding the blistering quality of *Plumbago rosea*. In the Raj-Nighant of Narhar Pandit, in the Bhav Prakash of Bhav Misra, in Madanpal-Nighant and in *Dhanvantari Nighant*, *Plumbago rosea* is spoken of as Anal or Vanhi, which means fire; and we know that when fire is brought into contact with human skin, it either burns the skin or blisters it. It produces a blister where it does not burn. Even so it is with the root of *Plumbaago rosea*. In sanskrit it is also called *Chitrak*, from whicxh we derive our Marathi name. Rendered into English *Chitrak* means the marker. It distinctly marks the skin.

The experiments which Dr. Dymock and his colleagues are reported to have made* with two cwts. of *bazar plumbago root* (the italics are mine - K.R.K.) are worthy of notice here. From these two cwts. was obtained half-a-fluid ounce of a deep yellow oily fluid having a peculiar penetrating odour. The effect of a few drops of this oil "smeared upon the upper part of the arm was not vesicating and occasioned no inconvenient symptoms." This points to one of two

*Pharmacographia Indica, vol. ii, p. 333

things-either that (1) the vesicating principle is volatile and that it had disappeared from the roots before they were experimented on, they being bazar-bought, and consequently perhaps of old standing, at any rate not quite so fresh as required; for all the previous writers have distinctly said that the vesicating property exists in the fresh root; or (2) that the vesicating principle is decomposed or chemicvally altered during the process of experimentation. This may be supposed to ba mere conjecture on my part; and so it is. It is only as such that I submit it for the consideration of the colleagues of Dr. Dymock- Dr. Warden and Mr. Hooper- who helped Dr. Dymock in bringing out the "Pharmacographia Indica," with the humble suggestion that in their future experiments fresh root of *Chitrak* may be tried. The bazar root though apparently fresh and moist, is necessarily old and, in consequence, worthless for the purpose of accurate analysis. It should be taken fresh from the ground. Besides I find that Dr. Dymock and his colleagues had "to dry the root before it could be powdered." ("Pharmacographia Indica," p. 334 , vol. II). It is necessary that further researches should be made without drying the root, for the blistering quality is supposed to exist in the roots of plants of the entire genus *Plumbago when they are fresh.* In quoting the researches made by Prof. Fluckiger and Mr. T.E. Gerock on *P. europaea*, Dr. Dymock observes that the Flumbagin which it contains, or which is obtained thereform, is "probably the product of a rapid oxidation of *Some primary substance contained in the plant.*" (The italics are mine - K.R.K.) May not this "*Primary substance,*" I ask, be the vesicating agent which in Dr. Dymock's experiments underwent rapid oxidation? May not the inertness of the oil he obtained from the root be due to such rapid oxidation as he conjectures? There is room for yet more accurate research in this direction. I have only to repeat what Dr. Dymock has himself said with regard to the action of the root of *Plumbago rosea.*" -"Its action is well worthy of accurate scientific investigation." ("Pharmacographia Indica," vol. II., p. 331). There is need for further observation as to what is the chemical constitution of the blistering agent in this root.

The root of *Plumbago europaea* has long been known in Europe, and is used for various purposes. It is employed by beggars, says Lindley ("Vegetable Kingdom" p. 641, 1847), "to raise ulcers upon their bodies to excite pity; its root is so acrid that it is used in Roumelia for causing issues, and even as a vesicant." A young woman who had applied it is said to have affirmed "that the pain it occasioned was intolerable and that she felt as if being flayed alive." It is also said to be emetic. The blistering properties of the entire genus *Plumbago* were not unknown to Pliny, for he says, observes Loudon,* that this plant (I should rather say the genus *Plumbago- K.R. K.) was so called "from Plumbum,* because it possessed the power of curing a disorder in the eyes called by that name, which appears to have been the same as cataract." There is also a modern reason for the application of the name to this genus. *P. europaea* is called toothwort and also dentelaire, Fr., from its curing the toothache, "for which purpose the bruised root is chewed when it excites by its causticity a healthy salivation, but stains the teeth a lead colour." The "Pharmacopoci of India" notes that the Javanese apply the root of *P. rosea* topically for the cure of toothache. Here is additional testimony from authoritative writers† :— "The Plumbagos contain a very caustic colouring matter; the root of the European species contains a fatty substance which gives a leaden colour to fingers and paper." In referring to *Plumbago europaea,* Beck‡, remarks as follows :- "Suavages observes that the workmen who make use of the decoction of this plant for the purpose of obtaining a yellow dye, are tormented by a severe headache if they work longer than six hours." As a stainer of the skin of a leaden hue, I may confidently say that our *Plumbagho rosea* has no such notoriety. Dr. Norman Chevers says that it is a very active but painful and a very uncertain vesicatory (The italics are mine - K.R.K.)

The root is used in various forms in India for criminal purposes. So far back as 1837, Dr. O'Shaughnessy noted this fact and succeeded in detecting the poison b chemical analysis as indicated in the foregoing remarks. Apart from its blistering property when applied to human skin, there is yet a more serious and dangerous use of the root *Plumbago rosea* criminal purposes which must seriously engage our attention. Dr. O'Shaughnessy says that it is very largely used as an irritant to occasion abortion. For this purpose "it is introduced

* Eucyclopædia of Plants, pp. 118-119, 1829.

† E. Le Maout and Decaisne's system of Botany, translated by Mrs. Hooker, p. 528 1873.

‡ Medical Jurisprudence, p. 838, 1836

into the vagina and applied directly to the neck of the uterus." To this observation of Dr. O'Shaughnessy, Dr. Norman Chevers adds a note* that the root of *Plumbago rosea* is frequently used with far more violence than is described by Dr. O'Shaughnessy. Referring to Dr. Norman Chevers remarks on criminal abortion[†], I find that Rai Bahadur Kannya Lall Dey affords him valuable information regarding the use of *Chitrak* (red variety) for the purpose of procuring abortion. In a document the Rai Bahadur handed over to Dr. Norman Chevers, as the doctor's work was goint to press, the veteraan Babu says in unmistakeable language that *lal Chitra* (the Begali name of *Lal chitrak* or *tambdi chitrak* as called in Western India), "is very certain in its action, producing shivering almost immediately after the application, followed by abortion in two or three hours; and it may be used at all periods of pregnancy, still the foetus is always expelled lifeless, and the woman is in great danger." Such is also the experience of Dr. Allen Webb. Apart from this directly local use of the root of *Plumbago rosea* on the generative organs, Dr. O'Shaughnessy has shown that it is administered internally, that is to say, by the stomach to cause abortion. In support of this Dr. Norman Chevers cites a fatal case mentioned to him by Kannya Lall Dey which occurred in 1857, and in which a mixture of *Chitrak* root and some arsenic was administered to a girl for the purpose of making her abort[‡]. The indigenous writers I have mentioned above, are totally silent as regards the abortifacient quality of the root of *Chitrak*. Chakadatta, as quoted by Udoy Chandra Dutt[§] is the only ancient Hindu authority who states that *Plumbago rosea* has a specific action on the uterus. "The root of *Plumbago rosea*" say he, "taken internally, will expel the foetus from the womb whether dead or alive." Dr. Lyon, as chemical Anaalyser to Government, Bombay, refers[||] to a case from Sangamner, in which "some pieces of stick, stated to have been used for the purpose of procuring abortion, were found to be armed at the end with cotton covered with a paste in which, on chemical examination, *Plumbagin*,

* Medical Jurisprudence p. 251, 1870.

† Medical Jurisprudence, pp. 712-19, 1870.

‡ Dr. Norman Chevers Medical Jurisprudence, p. 116, 1870

§ The Materia Medica of the Hindus, p. 187, 1877

| | Report for the year 1881-82, p. 11

the characteristic principle of *Plumbago rosea* and *Zeylanica*, was detected." In a subsequent annual report*, reference is made to another case of death after abortion which occurred at Satara, and in which *Plumbagin*, the characteristic principle of *Plumbago rosea* and *Zeylanica* was detected in a lump of paste found lying in the vagina of the deceased. In this case arsenic also was detected in minute quantity in the liver." Among the modern Hindu Writers who recognize the abortifacient properties of *Chitrak*, are Dr. Sakharam Arjun and Mr. Jaikisson Indraji.

Taken Internally, the root of *Plumbago rosea* acts as an irritant. Dr. Dymock states that when given internally in large doses, it acts as a narcotico-irritant poison. "In Small doses it acts first as a powerful stimulant of the mucous membrane of the digestive organs and after absorption, as a stimulant of the excretory glandular organs†." Colonel Drury observes that the root is not unfrequently used internally by the nastives in Bengal for criminal purposes. "Taken internally," says he, "it is an acrid stimulant and in large doses acts as an acro- narcotic‡." Dr. Norman Chevers says that the root is criminally used internally in poisoning *men.*§" He cites two well- recorded cases, from which I give the following extracts :-

(Case 1). In May, 1843, a woman living in Howrah, pounded a small piece of a reddish-coloured root which had been given to her, and having mixed it with milk, gave to her husband. He was taken ill in about two hours after swallowing it, and died after having vomited ten times and being purged once. The Civil Assistant Surgeon observed a slight flush of inflammation on the surface of the stomach and small intestines. * * * The surface of the stomach was found by Dr. Mount who examined it, to be much corrugated and covered with small patches of inflammatory blush. On analysing the contents of the stomach, some of the matter vomited, and the remains of the food said to have been taken by the deceased, Dr. Mount could detect no trace of any mineral poison, but obtained the distinct evidence of the characteristic appearance produced by *lal chitra*.

* Reports of the Chemical Analyser to Government, Bombay for the year 1885, p. 7

† Pharmacographia Indica, vol. ii, p. 331

‡ Useful Plants of India, 1873, p. 348

§ Medical Jurisprudence, 1870, p. 252

(Case 2). In October, 1853, the Civil Assistant Surgeon of Azimgur forwarded the stomach, with its contents, of a man supposed to have died by poison, together with a substance found near the body and supposed by the police to be poison. The stomach was received by Dr. Mount in an advanced state of decomposition, but still exhibited traces of inflammation and was found to contain *lal chitra.* The substance sent was also found to be a portion of the root of the *Plumbago rosea.*

Dr. T.D.C. Barry, Chemical Analyser to Government, Bombay in his annual report for 1891, refers to a case of poisoning by *Plumbago rosea.* The details of this case are not before me, nor have I seen the report. I am writing from the Government resolution on that report as published in the newspapers. I therefore refer the reader to the report itself for further information.

The table given by Dr. Dymock (pp. 335-339) in vol. II. of the "Pharmacographia Indica" shows the particulars of Plumbago poisoning in India, and is well worthy of the perusal of those who are interested in the toxicology of this plant and its active principle *Plumbagin.*

To summarize briefly what I have stated above, the deleterious effects of *Plumbago rosea* may be classified under three heads. :-

(1) Externally or locally in men and women as a blistering agent.

(2) Internally i.e., by the stomach in men and women as an acro-narcotic or narcotic-irritant poison.

(3) Internally, i.e., by the stomach, and locally through the genital organs as an abortifacient.

Reprinted from
Journal Bombay Natural History Society
Vol. 10(1) : 88-107. 1895

PART XIII.

ANACARDIUM OCCIDENTALE — (*Linn.*).
Natural Order -ANACARDIACEAE.
MARATHI — काजु (*Kaju*).

This is the *Cashew-nut* tree of English writers. It is quite an ornamental tree when in full blossom and fruit, displaying the most gorgeous colours in various shades. It is an evergreen, throwing out fresh foliage every year. From all the evidences I can gather, it is a tropical American and West-Indian plant naturalized in the Eastern tropical regions. In India, and especially in Western India, it is largely met with along the sea-coast and its adjacent sandy tracts, and in the rich *moorum* soil inland. It thrives very well in the Thana district.

It varies in height from tem to twenty feet. The tree blossoms once a year regularly, in the cold weather, even as early as November. The flowering often lasts till April and even later. The seed is ripe in May and June; not unfrequently as early as February.

Generally speaking, it is something short of a tree, and something more than a mere shrub.

TRUNK. — Short, thick, and regular up to three or four feet above the ground. Beyond that, it is crooked, and very irregular in branching. The branches, however, taken altogether, make it a very graceful plant, quite fit to be introduced into any garden in India which would like to display splendid colours in foliage, flower, and fruit. The crooked branching is no bar to its garden- beauty.

BARK. — Considerably rough. In old trees, says Sir William O'Shaughnessy, "it is deeply cracked." The juice from the stem is not milky, but thickish and resinous, of a slightly brownish tinge, blackening on exposure. "From the bark," says Baillon,* "oozes a yellowish hard resin, having almost the appearance of yellow amber, the Cashew gum (FR. *Gomme d'Anacarde*) soluble, and used for nearly the same purposes as gum-arabic." This shows that the resinous exudation from the bark does not possess any irritant properties. I may observe, however, that in the plants I have examined in the Thana district, I have not been able to notice any such yellow amber-coloured resinous exudations on the bark. There is a gummy exudation from the bark, but it is not resinous. Of this more hereafter. The bark contains much tannin. It is used to *curry* leather, and to prepare astringent lotions and gargles.

WOOD. —Dark brown; inferior in quality, as it is not generally deemed of any value in carpentry;† but major beddome says it is used for packing-cases; and Captain Dance says, it is, in Tavoy, a *large tree* used in boat-building, and it forms a charcoal which the iron-smiths there consider the best for their trade.

LEAVES. — Simple, smooth, alternate, exstipulate, petioled, quite entire, ovate or obovate; with a slightly rounded emarginate apex; smooth on both sides and of a hard texture; "Narrower but obtuse at the base." (Wight and Arnott.)§ The length of the leaves is from four to eight inches; the breadth from three to five inches.

The *Venation* of the leaves is generally well marked, whitish and prominent on the under-surface. The *nerves* generally in ten pairs, often less; nearly horizontal; sometimes bifurcating faintly. The leaves, like the bark, contain much tannin.

PETIOLE. — Varying in length from a quarter to half an inch; slightly grooved on its ventral aspect; at times almost cylindrical.

INFLORESCENCE. — In panicles which are bracteate, copious and situated at the apex of old branches bearing both male (*i.e.*, barren) and hermaphrodite flowers scattered without any special order on the pedicles. Note that the hermaphrodite flowers appear and develop into fruit, long before the new foliage of the year is thrown out. Sir William O'Shaughnessy observes that "There may be trees which

* Natural History of Plants, Vol. v, page 303.

† Balfour's Timber Trees of India, page 16, 1870.

§ Prodromus Flor. Ind., Vol. i., page 168

produce barren flowers only." I am not quite sure of this, so far as my observation goes with regard to plants growing in the Thana district. I am quite aware of the fact, however, that some of the branches of the panicles may have both barren and hermaphrodite flowers, and yet bear no fruit, although their richest blossom may apparently hold forth the hope of a mature fruit and seed.

PANICLES. — "Corymbose, branched and spreading" (Wight and Arnott)*; six to ten inches long; pubescent. The branches of the panicles are long and naked to the tips when the flowers are collected (Hooker).

BRACTS. — Leafy, numerous, lanceolate, hoary. Hooker† says that they are "*gibbous*." This does not appear to be the case on this side of India. (K.R.K.)

BRACTEOLES — Present at the base of pedicels; "broadly ovate," say Wight and Arnott (*op. cit.*), generally lanceolate; acuminate.

FLOWERS. — Small; pentamerous; polygamous, i.e., a flower-stalk may bear both male or barren, and hermaphrodite flowers. The diameter of the flowers is generally about 1/3 inch. Their colour is pale yellow with pink longitudinal stripes, often of a deep crimson colour on this side of India.

The odour of the flowers is strikingly penetrating even from a distance, and partakes of the characteristic mixed odour of cloves and cinnamon. There is no tree or shrub on this side of India which is so easily recognizable from its mere odour as this plant. Given this odour, there is sure to be a Cashew-nut plant in flower close by. Such has been my experience each time I have been in the close vicinity of a Cashew nut plant in flower. The odour is quite unmistakable. Some say the odour is disagreeable, but opinions differ.

ÆSTIVATION. — Of Calyx and Corolla, imbricate.

CALYX. — Inferior, cleft nearly to the base.

Sepals — Erect; deciduous; "Oblong, conic, acute and pretty smooth" (Sir William O'Shaughnessy)§. The base of sepals accrescent, forming an erect disk.

---

* Prodromus Flor. Ind. vol. i., page 168.

† Flora of British India, vol. ii, page 8

§ Bengal Dispensatory, page 280, edition 1841

COROLLA. — Alternate, with the sepals free, entire, acutely re-curved.

Petals. — Acuminated, linear-lanceolate, twice as long as the sepals; often longer.

Note that the sepals and petals remain unchanged after flowering (Kurz).*

STAMENS. — Usually nine; all fertile. One of these is nearly twice as long as the rest; the length of the remaining stamens, however, is by no means uniform; some are smaller than others, often alternately so. Baillon notes (*op. cit.*) that "They are all verticillate, 2-seriate."

FILAMENTS. — Connate at the base, free upwards; adnate to the disk, "Which is prolonged upwards," says Baillon, "in the shape of a glandular ring." It is on this disk that the petals are situated†.

ANTHERS. — 2-celled; introrse; longitudinally 2-rimose; basifixed. The anther of the longest filament is twice as large as that of the minor ones. It is of a bright crimson colour. The anthers of the minor filaments vary in colour from orange-yellow to crimson on this side of India.

PISTIL. — In the male flowers minute, with a very short style. In the hermaphrodite flowers distinct and well-developed.

OVARY. —In the hermaphrodite flower free; campylotropous; superior§; one-celled; obovoid or obcordate. Baillon describes it as "compresso-obovate, or obcordate," hence gibbous. This is a more accurate description I think.

STYLE. —Simple or solitary, corresponding to the one-celled nature of the ovary; filiform; excentric; becoming convolute, as if to bring the simple stigma in contact with the large anther of the long filament (Roxburgh).

STIGMA. — Minute; often tinged crimson.

OVULE. - Solitary, long, conical, inserted at the summit of a sub-erect ascending funicle. The funicle is thus lateral. The ovule, says Baillon (*op. cit.*), is the same as that of its congener Mango, at

* Forest flora of British Burma, vol. i, page 302.

† Wight and Arnott's Prodromus Floræ Indicæ, vol. i, page 168.

§ Hooker puts a query against this, it is not known why. (*Vide* Flora British India, vol. ii, page 8) – K.R.K.

the base, though finally the funicle supporting it is inserted a little higher.

CHALAZA. — Superior.

MICROPYLE. — Introrse, inferior near funicle.

FRUIT. — An ash-coloured nut, and as such dry and indehiscent; kidney shaped; usually about an inch long, sometimes a little bit longer; a quarter of an inch broad at the hilum with fully developed; somewhat compressed; marked at the hilum with umbilicus, and the cicatrix of the style. Mesocarp soft, corky, lacunose and oleo-resinous. The epicarp, or pericarp, as Sir William O'Shaughnessy describes it, is "coriaceous," and smooth (K.R.K.); and not "woody" as Baillon says.

The fruit is termed "Nucumentaceous" by Baillon. It is seated on a succulent obovoid accrescent receptacle formed from the enlarged disk and top of the peduncle of the fecundated hermaphrodite flower. It is wrongly called pyriform. It is anything but pyriform, in my humble opinion. So far as I know, it is more obovoid than pyriform. This obovoid receptacle develops to the length of from 2 to 4 inches, being sometimes two inches at its broadest part. The peduncle is narrow, cylindrical, and rather hard before the fecundation of the flower; after fecundation it thickens, becomes succulent, and assumes an obovoid form. This thickened fleshy peduncle is ordinarily known as the Kaju fruit, though the real fruit, botanically speaking is the Kidney-shaped body of the termination of the fleshy peduncle. This fleshy peduncle is not unoften of an irregularly oval form. O'Shaughnessy in his Bengal Dispensatory says it is "oval."

SEED. — Kidney- shaped, ascending, corresponding to the shape of the pericarp.

TESTA. —Membranous but crisp, and easily removable; of mottled reddish-brown colour outside, and deep crimson inside. It is of an astringent aromatic taste. Observe that Hooker says that the "testa is adherent," but it is not so in reality. It is separable by a "resinous fracture" with the merest tough or crush between the fingers, exposing to view the cream-white or milk-white cotyledons. The testa under such circumstances cannot be in any way said to be "adherent."

COTYLEDONS. — Glossy white; sometimes cream-coloured; thick, softly fleshy, with a plano-convex exterior, and a deeply grooved interior; distinctly kidney-shaped, answering the shape of the pericarp generally.

RADICLE. — Short; exserted; hooked, or "incurved and inferior," as Baillon describes it. I think Wight and Arnott describe it more accurately when they say it is curved upwards from the base of the cotyledons.

ALBUMENS. — None.

EMBRYO. — Having the form of the seed; erect, milky-white; *Plumule* minute, milky-white.

REMARKS.

The following are the synonyms of the plant —

1. *Acajuba occidentalis* (Gaertner De Fructibus, vol. i. pp. 192-193).

2. *Cassuvium pomiferum Lam* (Rheede Mal. iii. t. 54). *Rumphius Amb.* i, p. 177; 69).

3. *Acajou* — South America.

The species known as *Anacardium indicum* and referred to in Johnson's Gardener's Dictionary recently edited by Wight and Dewar (1894), evidently from DeCandolle's "Prodromus"* is not different from the *Anacardium occidentale* in any way. Paxton† also mentions a separate Indian species under the same name, *viz, Anacardium Indicum.* From the description given by the elder DeCandolle, there does not seem to be much difference between the true original American species described as *Anacardium occidentale,* and the naturalized oriental plant separately described as *Anacardium Indicum* This will be fully explained in detail later on. That the two species separately described by the elder DeCandolle are one and the same has been amply proved by the younger DeCandolle, as will be seen from latest work of the latter, issued in connection with the "International Scientific Series" under the auspices of Ms. Kegan. Paul. Trench & Co., entitled "Origin of Cultivated Plants" (1884). In this rich mine botanic lore, the following are his mature and deliberate opinions

* Prodromus Syst. Nat. Regni Vegetabili, Pt. II, p. 62.

† Botanical Dictionary, p. 31, 1868, p. 31. New Edition by Samuel Hereman.

regarding the identity of *Anacardium occidentale*, var. *Americanum*, of the Western tropics with *Anacardium occidentale*, var. *Indicum* of the Eastern tropics. "The most erroneous assertions," says Alphonse DeCandolle, "about the origin of this species were formerly made, and in spite of what I said on the subject in 1855,* find them occasionally reproduced." In proof of this he mentions Tussac, who says† that *Anacardium occidentale* "is an East Indian species," thus aggravating Linnaeus' mistake, who believed it to be Asiatic and American. Divorced, as I am in India, from the full literature on this subject, especially continental, including the Journal referred to by A. DeCandolle, in the foot-note quoted below, and marked 1 at page 199 of his "Origin of Cultivated Plants," I am unable to examine the arguments he has used to show that there is no such thing as two distinct varieties, *viz.*, an American and an Indian, of the species *Anacardium occidentale*, as described by his illustrious father Pyrame DeCandolle in his voluminous "Prodromus." I am in a position, however, to unhesitatingly accept the conclusions of Alphonse DeCandolle, the son, in preference to those of Pyrame DeCandolle, the father, as the former are the result of a wider acquaintance with geographical distribution of the Cashew-nut plant.

But before I proceed to quote here fully the grounds on which I accept Alphonse DeCandolle's conclusions, I may note one point with reference to his remark (*vide p.* 199 *op. cit.*) which runs thus — "The French name *Pommier d'acajon* (Mahogany apple tree) is as absurd as it is possible to be. It is a tree belonging to the order *Terebintaceae or Anacardiaceae*, very different from the *Rosaceae* and *Meliaceae* to which the apple and the mahogany belong; the edible part is more like a pear than an apple, and, botanically speaking, is not a fruit but a receptacle or support of the fruit which *resembles a large bean.*" The italics are mine. The true fruit neither resembles a "bean," nor can it be called a "large bean." From my foregoing remakrs, it will have been apparent that I am not disposed to term the fleshy development of the peduncle, called in popular parlance the *Kaji fruit*, "a fruit like an apple." Nor would I, with

* Geog. Bot. Rais, p. 873 from the foot-note of A. DeCandolle in his Origin of Cultivated Plants - K.R.K.

† Flor. des Antiles, III, p. 55.

equal reason, be disposed to agree with A. DeCandolle in calling it a "fruit like a pear." In shape and botanical structure it is neither like an apple, nor like a pear. But if I were one of those who in former years were disposed to call it an apple, I would with equal vigour and no less reason retort in the very words of Alphonse DeCandolle, that to say that the succulent development of the peduncle "resembles a large bean" is "as absurd as it is possible to be." No tree of the order *Anacardiaceae* has any pretensions to be called "*bean-like*" any more than "*apple-like*" or "*pear-like*."

But to proceed to the grounds on which A. DeCandolle considers that the *Anacardium occidentale* is an introduced plant in the Eastern tropics, I may at starting say that his statement on this subject, is conclusive, and may well be quoted *in extenso*. I shall, however, add my own observations wherever necessary.

"The species *Anacardium occidentale*," says he, "is certainly wild in the forests of tropical America, and indeed occupies a wide area in that region; it is found, for example, in Brazil, Guiana, the Isthmus of Panama, and the West Indies. Dr. Ernest believes it is only indigenous in the basin of the Amazon River, although he had also seen it in Cuba, Panama, Eucador, and New Granada. His opinion is founded upon the absence of all mention of the plant in Spanish authors of the time of the Conquest — a negative proof which establishes a mere probability."

Whatever the question may be as regards the geographical distribution of the Cashew-nut plant in the Western world, there is no question as to its prevalence in the Eastern tropics, as a mere naturalized exotic. It is certainly not wild in the Konkan, although A. DeCandolle says that both Rheede and Rumphius "indicate" this plant in the south of Asia. According to Rheede (Malabar, iii, pl. 54) the plant is common enough on the Malabar Coast. Rumphius has "different historical and philological proofs."* "which," says Alphonse DeCandolle, "have convinced me that its origin is not Asiatic." I think I can safely vouch for this statement, for so far as I am albe to ascertain, I find no Sanskrit writer of ancient date referring to this plant. Let me add to this the testimony of Rumphius who, as A. DeCandolle says, is always accurate. Rumphius spoke of an

* Rumphius "Herb Amboin". I, pp. 177, 178.

ancient introduction by the Portuguese into the Malay Archipelago from America. The Malay name he gives, *Cadju*, is American; that used at Amboyna means "*Portugal fruit*;" that of Macassar was taken from the resemblance of the fruit to that of the *Jambosa*. Rumphius says that the species "was not widely diffused in the islands," in his days. "Garcia ab Orto did not find it in Goa in 1550, but Acosta afterwards saw it in Cochin, and the Portuguese propagated it in India and the Malay Archipelago."*

I have already observed that in describing the species *Anacardium occidentale*, the elder DeCandolle mentions two varieties, namely, (1) *Americanum*; (2) *Indicum*. With reference to this the young DeCandolle observes that "the existence of the same arborescent species in Asia and America was so little porbable that it was at first suspected that there was a difference of species or at least of variety, but this was not confirmed." This will be made clearer by giving below a translation of the description†, originally given in Latin, of the two varieties. It is as follows —

(a) *The American variety* has a thickened peduncle, about ten times larger than the nut, the longer filament having an abortive anter (*anther orbato*), and being slightly dilated at the apex.

(b.) *The Indian variety* has a thickened peduncle scarcely three times longer than the nut, the longer filament being endowed with a thick anther (*anther crassa donato*), other anthers being slightly dwarfed (or ill-developed) (*coeteris subeffoetis*).

It would appear from the above details that the differences are chiefly in the development of the fleshy peduncle in relation to the nut, and in the abortive or thickened condition of the anther-lobe. Judging from the various developmental peculiarities of the anthers in the Indian flowers of this plant, one could hardly consider any chance-arrest of development, or chance-increase in the size of the anther-lobes a sufficient justification for the creation of a new species or even a permanent variety.

Apropos of Alphonse DeCandole's remarks on the historical aspects of this question, let me here briefly refer to the latest utterances of an erudite Botanist, Dr. Henry Trimen, F.R.S., Director of the

* A. DeCandolle's origin of Cultivated Plants, pp. 199-200.

† Pyrame DeCandolle, Prodromus Syst. Nat. Regni Veget. Pt. II, p. 62.

Royal Botanic Gardens of Peradynia, in his Hand Book of the Ceylon Flora,* issued four years short of a century and a half after Carolus Linnaeus — the immortal Father of Universal Botany—gave to the world his *Systema Florae Zeylanica* from Upsala. Linnaeus in 1749 spoke of the Cashew nut plant as "arbor hodie botanicis notissima." If the plant was so in the days of Linnaeus, it is no less so in our own day. It has no parallel in the vegetable kingdom. Dr. Trimen observes, with reference to the prevalence of this plant in Ceylon, "that the Cashew-nut is so completely established in the low country, especially in the sandy ground near the sea, as to have all the look of a native tree. * * * * The Sinhalese call it *Caju*, an adoption of the name used by the Portuguese, who no doubt introduced it. The Tamil name for the edible part is '*Montirikay*†.' It is figured in Beddome's Flora Sylvatica, t. 163. Its native country is Tropical America, and '*Acaju*' is the Brazilian name."

"According to Blume and Miquel," says A. DeCandolle, "the species is only cultivated in Java." It is known there under the name of *Jamboc monjet*. It is frequently found in the gardens of Java, where, however, it is smaller than in Surinam, as its height in Java does not exceed that of an apple tree in Europe‡.

The Cashew-nut is not unknown in Singapore. Its Malayan name there is *Jambo irong*. Its fruit is said to be coarse, and not eaten there. Its nut is said to be astringent§. As regards the occurrence of this plant in Burma, from the fact that it is found only "in the beach forests and along sandy sea-shores and dunes of Chittagong and Tenas-serim||," and also from the fact that it is only a cultivated plant in villages, it would appear that it is an introduced plant.

But to return to Alphonse DeCandolle's "Origin of Cultivated Plants" once more. He observes that although Rheede truly says that the Cashew-nut is abundant on the Malabar-Coast, Rheede only quotes one name which seems to be Indian — *kapa mava*; all others are derived from the American name.

---

*Part I, pp. 73-74 1893.

† In India it is pronounced as *Mundirikai* (*vide* p. 72, "Five Hundred Indian Plants," by C. Stolz, 1891, Managalore Basel Mission Book and Tract Depository). It is known as *Gerumava* in Dharwar, says Talbot (*vide* his Systematic List, 1894, p. 61).

‡ *Flore et Pomone de java*, par Madame Berthe Hoola van Nooten, Brussels, 1866.

§ Our Tropical possessions in malayan India, John Cameron, Esq., F.R.G.S., Appendix I, p. 399, 1865.

|| Kurz, Forest Flora British Burma, vol. I, p. 311.

I have, I think, sufficiently indicated in the foregoing remarks that the Cashew-nut plant is an exotic naturalized in the entire Eastern tropics. As already observed, the ancient Sanskrit writers make no mention of it. In the sanskrit-French Dictionary of Burnouf and Leupol (Paris, 1856), and in the still more recent Sanskrit Worter-buch of Otto Bothlingk (*Petersburg*, 1879), no Sanskrit equivalent of *Anacardium occidentale* is to be found. Both these dictionaries are replete with names of Indian plants with their botanical equivalents. The plant is not mentioned in Raj-Nighant, nor is it to be found in Madan Pal-Nighant. There is no reference to it in Dhanvantari-Naghanta either. The only reference to it in Sanskrit that I find is a solitary *shloka* quoted by three different compilers. The works of these compilers are of very recent dates, namely — (1) *Nighant Ratnakar*, published in 1867 (Bombay); (2) *Guanadosha Prakash*, Published in 1892 (Poona); (3) *Nighanta Sangrah*, published in 1893 at Junagadh, by Raghunath Indraji *alias* Kata Bhat. This learned writer, in citing a passage mentioning some of the qualities of the plant, refers the passage to *Bhav Prakash*; but I am at the present moment unable to verify this quotation. Even supposing the passage is to be found in *Bhav Prakash*, it being a recent work, it would in no way militate against the fact that the Cashew-nut plant is in India an introduced plant.

Both Mr. T.N. Mukarji and the Rev. Mr. A.K. Nairne make the mistake of saying that the fleshy peduncle when fully developed is of only one colour. "It is of a bright scarlet colour," says the former;* "the fruit is red," says the latter†. As a matter of fact it is not only scarlet and red, but orange and lemon-coloured, with all the shades between. The tint ranges from light canary to the brightest scarlet one can imagine under the dazzling sun of the Indian sky. Baillon has said that "the colour is white, yellow, or red," according to variety. I have not seen a white *Kaju* in this district; I may observe, however, that the colour, whatever it may be, is not determined *according to variety*, as Baillon supposes, for, on one end the same plant one may find the fully-developed fleshy peduncle in all shades of colour; often so, one & same branch. The colour is mere matter of accident,

* Handbook of Indian Products, p. 74, 1883.

† The flowering Plants of Western India, p. 68, 1894.

undetermined by variety. Balfour says that the fleshy peduncle is pea-shaped; he evidently means pear-shaped. The juice is described by him as *very acrid*, and highly recommended in dropsical habits. As observed before, ideas differ with regard to the acrid nature or otherwise of the juice of the fully-developed fleshy peduncle. Some use the so-called fruit without much suffering; others find it irritating and will not use it. Nairne observes that "though tempting to the eye, it is *excruciating* to the palate, though Oliver calls it edible." "The flavour," says Baillon, "is sourish or more or less acrid and astringent. Conserves are made from it, and particularly that famous *Fools, Confection* of Hoffmann the use of which, it was said, gave intelligence and memory to those most destitute of them." I am not aware of any such benign and beneficent property in the so-called fruit of kaju, else our village boys, who, notwithstanding its acridity, so largely and with such avidity use it throughout the period when the fruit is in season, — our village boys, I say, would be giants of intellect, proportionate with the quantity of fruit they consume! With regard to its use in Java, Madame Nooten says that "the fruit has a very tempting colour, but this fruit, so attractive to the eye, is not eatable." "It is true," she adds, "that the edible part is sometimes eaten by the Javanese, but they find few imitators on account of the sharp astringency of the juice which possesses a virulent principle." What that virulent principle is nobody has found out, so far as the acrid taste of the so-called fruit is concerned. I remember to have eaten the fruit muself as a boy with relish, though not without suffering from irritation of the throat lasting for some time after eating it. And I have not failed to appreciate it since. Though acrid, it is slightly sour, generally sweet, often *very* sweet, and having an aromatic odour which some may not like. A pinchful of common salt added to the sliced fruit, when eating, reduces the acridity. Baillon says that fermented drinks - wine, alcohol, and vinegar - may be extracted from it. "In Brazil," says he, "it is called the Sarsaparilla of the poor, and used as sudorific, diuretic and antisyphilitic." Through the good offices of Mr. M.C. Pereira of Bandra, I have been able to obtain for purely medicinal purposes a small quantity of a clear spirit (alcohol) obtained from the so-called kaju fruit. It was manufactured at Goa whence Mr. Pereira obtained a small quantity of it for medical purposes. It is seldom if ever used internally. It has a sweetish aromatic taste, partaking of the flavour of the mature fruit; and it goes under the name of "*Kajuchi-Daru.*"

With regard to the nuts of the Cashew, I have come across some very strange passages in a work on the Folklore of Plants by Richard Folkard. He says* that the nuts are supposed by the West Indians to excite the passions. The later Hindu writers I have mentioned above recognize some such property in general terms. "The Negroes of the West Indies say," observes Folkard, "that a branch of the Cashew-nut tree supplied the crown of thorns used at our Saviours' crucifixion, and that, in consequence, one of the bright golden petals of the flower became black and bloodstained." The spirit of profanity "could no further go!" Well may I exclaim so. Judging from the circumstance that the crucifixion of Christ occurred several centuries before the plant was introduced into the Eastern tropics from the Western World, and noting the stern botanical fact that the flower of the Cashew-nut plant has neither "a bright golden petal" nor a black stain thereon, albeit the fact that its petals have crimson streaks, call them "blood-stained" if you will, Folkard's reference must be taken to be an utter myth, the product of a fertile but vitiated brain on the wrong side of the border-land of insanity. The plant has luckily no thorns to have any such unhallowed claim to have formed a part of the cruel and cursed "crown of thorns" that touched the brow of a saintly being and superbly sacred person in the last agonies of his earthly career.

Madame Nooten mistakingly supposes that a resin is obtained from the kernel. She is not a botanist, and is only writing from second-hand information. But she evidently refers to the resin from the pericarp. She is, however, right in saying that the resin "furnishes a more efficacious vesicatory than the Spanish fly or cantharides." But of this more later on. "The bark," says she, "contains a gum sold in Europe under the name of 'Cashew,' which," she adds, "is said to be a preservative against white ants." I think Madame Nooten is rather "mixed up" here. For, as Kanny Lal Dey, of Calcutta, says, it is the oleo-resin from the mesocarp that is applied to floors and refters of houses to prevent the attacks of white ants and other

* Plant Lore Legends and Lyrics, p. 271, London, 1892.

insects. The gum from the bark which Madame refers to is quite different in kind. It possesses no insecticide properties. It is free from the acridity which strongly characterizes the oleo-resinous product from the mesocarp of the nut. The gum from the bark consists chiefly of bassorine, common gum and gallic acid. It swells to a considerable size, but does not dissolve in water.* Balfour says "it forms a good substitute for gum arabic. In South America, bookbinders wash books with a solution of the gum in order to keep away the moths and ants." If it be a substitute for gum arabic, which has no insecticide properties, I cannot understand how it can keep off moths and ants. On the contrary ordinary gum arabic invites some kinds of ants. "The gum," says Balfour†, "should be collected when the sap is rising; 5-12 lbs. can be collected annually. Observed that Balfour further notes that the seeds are used to flavour Madeira wine.

There is a very strange series of remarks in Mr. T.N. Mukarji's "Hand Book of Indian Products" (*vide* p. 74), which must not be allowed to pass unnoticed here. They are all the more strange because he is supposed to know what he is writing about. "The kernel," says he, "when raw is *exceedingly acrid*" (the italics are mine — K.R.K.); "but when boiled it forms a delicious article of food." Nothing of the kind. There is not the slightest acridity about the unboiled "Kernel." Boiled or unboiled, it is equally delicious. True, some would consider the boiled or roasted kernel *more delicious*. Further, Mr. Mukarji says that "the oil of the *nut* is used as an anaesthetic in leprosy with advantage." Note that Mr. Mukarji is not a medical man of any pretensions. He is a layman. Let him understand that it is not the *oil of the nut*, but the *oleo- resinous product from the mesocarp of the nut*, or the true botanical fruit that is used in leprosy; and that too, not as an anaesthetic, but as an acknowledged remedy against the anaesthetic form of leprosy — *ergo*, as a cure for *anaesthesia*. It is strange that an oriental writer, knowing the plant well, should fall into such errors. He is merely a compiler.

As Rai Bahadur Kanny Lal Dey truly observes in his "Indigenous Drugs of India," the oil extracted from the kernel "by expression is

* O'Shaughnessy's Bengal Dispensatory, p. 281

† Timber Trees of India, p. 19.

exactly similar to olive oil, sweet and agreeable." I may add that it is free from any of the slighest acridity whatsoever. It is nutrient and emollient. Its specific gravity is said to be 0.9160. It is soluble in ether, and partially in alcohol. the seed gets rancid which left over a rainy season. It loses its bland taste. To obtain good oil, the seed must be taken fresh. The black oleo-resin from the mesocarp is alone acrid. It is quite different from the oil of the kernel. Balfour goes to the extent of even saying that oil from the kernel is *Superior* to olive oil or almond oil. I am prepared to endorse this view to the fullest extent. The sugared kernel knwon in the Bombay native sweetmeat-sellers' shops and to our children as *Kaju gola* is a familiar article in our local "fairs." Children visiting such "fairs," which are not uncommon in Bombay throughout the year on "full moon" or "no moon" nights,* are only too glad to invest their pice in purchasing sugared *Kaju gola* as their desired *bon bouche*, among other things, on such festive visits. They would not eat the sugared kernels of Kaju, if they possessed the slightest trace of acridity, no matter how much sugar might be used to cover it. The tender and highly sensitive throats of children would at once rebel against the use of an acrid sweetmeat under the disguise of sugar.

But to pass on from the kernel to the plumule. It is not very easy for me to understand how it is that Gaertner says that the *plumule is absent* from the seed of the cashew-nut†. He describes the plumule as the first bud or "gemma" of the new plant. Further, under the head of *Simple buds*, which he describes as having sessile leaflets, he says that "they are opposite and ovate-acuminate in the *Anacardium, Sitodium, Corylium.*" It is difficult to reconcile these two statements. Besides, from a personal examination of the seed, I find that there is a plumule, and a well-marked one too, answering the description Gaertner has given in his General Introduction, namely, "having sessile leaflets;" although in the body of his work (vol. I, p. 193), in describing the fruit of *Anacardium occidentale*, he asserts that *there is no plumule.* If the plumule is said to exist in the genus, it must be taken to be existent in all the species thereof, unless specially excepted. No such

* The *full moon* night in *Indian lunar months* falls at the end of the first half of the lunar month; the *no moon* night falls at the end of the second half of the same month - K.R.K.

† *Vide* p. CLXVIII, General Introduction, De Fruct et Sem., Vol. I.

exception seems to be made by Gaertner in either his General Introduction or his particular description of the Kaju plant. As a matter of fact the plumule does exist, although I quite admit it is likely to be absent from ill-developed seeds. Such a contingency is very likely to occur where exotic plants are considered and examined in all stages of development, possibly non-development of special organs, under skies where they do not grow, and where the examiner has to trust to dried specimens, in all stages of decay, and ravages of insects.

POISONOUS PROPERTIES

The poisonous properties of the plant exist in the acrid oleo- resin found in the mesorcarp of the nut. Its composition is said to be the same as that of the oleo-resin found in a mesocarp of its congener —*Semecarpus anacardium* described in a former contribution to this Journal.* It contains cardol and a fatty acid called anacardic acid. Dr. Dragendorff, Professor of Pharmacy in the University of Dorpat, Russia, distinctly observes that cardol possesses powerful vesicant properties which are not shared by anacardic acid. "Cardol is a colourless oil, accompanying anacardic acid," says Dr. Dragendorff†. "It is soluble in alcohol and ether, but not in water. It can be removed from suspension in water by agitation with chloroform. Contact with dilute potash for a short time does not result in the loss of the vesicant property of cardol, as is the case, when the alkali is concentrated, and the action prolonged. The observation of the Rev. Mr. Nairne that the shell of the nut contains a very acrid oil, from which anacardic acid is made, is not sufficiently accurate. In the article communicated by Mrs. Ida R. Brigham to the "London Pharmaceutical Journal and Transactions‡", I find that, according to Professor Staedler, in addition to cardol and anacardic acid, the oleo-resin of the Cashew-nut contains "a little of an ammonium compound." What this compound is, is not mentioned. Staedler, however, assures us that it is not irritant, nor is anacardic acid irritant according to him. Cardol has a "faint pleasant odour when warmed." It is, in my opinion, anything but pleasant. It is positively irritating — a fact which will be easily remembered by

* *Vide* Vol. IX, pp. 235-258.

† Plants Analysis - Qualitative and Quantitative. Translated, from the German by H.G. Greenish, F.I.C., London, p. 146, 1884.

‡ P. 708, vol. XII, 1881-82, 3rd series

those who have ever found themselves in the close vicinity of a place where the nut is being roasted. Dr. Dymock and his colleagues quote* the observation of Buchheim who says that the oleo-resin has a very faint and hardly acrid taste, and that three or four drops of it can be swallowed without marked effect. They also refer to Dr. Brassac's opinion regarding the quality of the oleo-resin in the following terms — "Dr. Brassac considers it to be a good, rapid and safe vesicant, producing a copious flow of serum, and notable reduction of hypertrophy in tubercular leprosy." The cases which I quote below will show that it is not always a safe vesicant in some healthy men, as it produces much constitutional disturbance and affects healthy parts of the human body far removed from the places which originally came in contact with the oleo-resin. This proves the volatile nature of cardol. There is no doubt that it irritaates the healthy skin often to a dangerous degree, producing not only vesication but regular eczematous inflammation of the skin. The anaesthetic skin of the leper is unable to appreciate the irritant property. "Cardol," says Dr. Dymock, "seems to be excreted with the urine, but partially also with faeces." This shows that cardol is a stable compound. The oleo- resin leaves a stain on linen, but it does not appear that the stain has the permanency so characteristic of the marking-nut. The oleo-resinous liquid, which is of a dark brown colour, contains 90 per cent. of anacardic acid and 10 per cent. of cardol. Dr. Lyon says that "the Cashew-nut juice," meaning, I presume, the oleo-resin, "appears to be seldom, if ever, used criminally in India." Dr. Newton, however, in the "Pharmacopoeia of India†" says that native women use the juice to produce abortion, applying it to the *os uteri.*"

Recorded evidence, however, is sufficiently strong to prove the vesicating properties of the oleo-resin. The Rev. Mr. Nairne seems to doubt this. Observe what he says‡ in quoting Kingsley's remarks to the effect that the fumes of the acrid oil "will blister the face if the cook bends over the fire." Mr. Nairne emphatically says, "I never heard of this difficulty in India." Mr. Nairne may not have heard of it but the belief is common on this side of India, and not unfounded on actual experience, that process of roasting the cashew-nuts is

---

* Pharmacographia Indica, vol. I, p. 387

† Medical Jursiprudence for India, p. 192, 1889.

‡ The Flowering Plants of India, p. 68.

not always free from danger. Where the nut is being roasted, its acrid, irritating fumes are unmistakably noticeable. Popular belief goes still further. Some think that the penetrating odour has the power to kill even horses. This, however, must be considered an exaggerated idea of the acridity of the vapour emanating from nuts undergoing the process of roasting. The oleo-resin has the reputation of being a vermicide, as well as an insecticide, which latter property has already been referred to.

Dr. W. Hamilton* says that the oleo-resin is of such a causticity as to blister the lips of those who incautiously suffer it to approach them; on this account these nuts are never eaten till after they have been well-roasted to dissipate the oil from the pericarp. After this, they may be used with impunity.

To test the action of the oleo-resin, Mrs. Brigham says, Professor Staedler "spread it over about one square inch of the surface of the breast, and covered it with a piece of blotting paper, also moistened with it. In the course of fifteen minutes a slight burning was perceptible, which rapidly increased, and attained its greatest energy in half an hour. The skin beneath the paper had become whitish surrounded by a red circle. The paper was allowed to remain in contact with the skin for three hours, which was then covered with thin vesicles that increased in size during the night, without, however, attaining the size of those usually produced by canthartides"; and yet we have the testimony of Madame Nooten, as already noted, that the oleo-resin (which she wrongly says is obtained from the *Kernel*) furnishes "a more efficacious vesicatory than the Spanish fly or cantharides." The affected part in Professor Staedler's experiment was not only very slow in healing, but the after-effects continued for a long time. "A second experiment made with the liquid, which had been treated with dilute hydrochloric acid, afforded the same result."

Mrs. Brigham adds that "Henry W Worthington states that he could fully corroborate the statement of writers concerning the irritant properties of the oleo-resin of the Cashew-nut, having himself seriously suffered from its effects while extracting it from bruised nuts by ether and subsequent evaporation. The United States Dispensatory notes a case of lady who was exposed to the

* Pharm. Journal, vol. V, p. 269, 1845-46.

fumes of the nut while roasting. Her face was so swollen that not a feature was discernible for several days. A similar case occurred in a boy, who cut open one of the nuts, eating a small portion raw, and by handling it, spread the juice over the different parts of the body. "The tongue, face, neck, hands, forearms, & c., were enormously swollen, red and very painful." In this very able thesis which Mrs. Brigham presented to the Massachusetts College of Pharmacy in 1878, she further states how she and her husband, both ignorant of the poisonous quality of the nut-shell, suffered very seriously from skin eruption while incautiously attempting to break the shell with their teeth. While attempting to extract the vesicating principles, Mrs. Brigham, in spite of the elaborate precautions of encasing her hands with rubber and of using instruments, suffered from vesication of her right check, whereon a small particle of the juice found its way. The result was that instantly a slight burning was felt. "As quickly as possible," she says, "the poison was washed away with water, and to prevent any minute particle from adhering, again washed with ether." But the mischief was done, and she suffered. It is hardly necessary for me to enter here fully into the details of what the fair experimentor had to pass through for the next eleven days and more. Suffice it to say that vesication was the result. The face was so disfigured "as to be unrecognizable to the most intimate friends." Hands, fingers included, suffered from a vesicular eruption attended with oedema; other parts of the body suffered in their wake. Mrs. Brigham closes her very interesting thesis with the warning words of science. These words will not suffer, I am sure, in being reproduced here as the guiding voice of bitter and painful personal experience. "Several authors," says she, "speak of its use (meaning the use of the oleo-resin - K.R.K.) for the removal of warts, corns and callosities, and some as a vesicant instead of cantharides. Considering the almost *infinitestimal* quantity of the poison which could have been absorbed in my personal attack and the gravity of the results, it would seem, in my judgment, to be unwarranted to apply a remedy so uncontrollable in cases where many other standard remedies might be used, which are not open to such objection." I am not disposed to underrate the value the painful personal narrative, but it appears these graphically described sufferings the husband and wife are singular in the extreme. They are by no means, however, what may be considered "unexpected." Some may escape the virulent action of a violent poison. It is best, however, from a scientific point of view, to recognize the possibility of serious suffering from the incautious use or handling, even if it were for experimental purposes, of the Cashew-nut oleo-resin.

Let me now proceed to the consideration of a case of poisoning by the Cashew-nut, reported in an American Journal* edited by Horatio C. Wood and Robert Meade Smith. It is stated in this journal that "In 1881 a gentleman connected with the United States National Museum while organizing the Materia Medica collection, came across some nuts unlabelled. He sawed one of them through to make a specimen, and got some of the tar-like substance between the shell and kernel on his hands." What I would say is — Would-be-arrangers of Materia Medica Museums beware! for see what happened thereafter. "A few days after†, the face and hands became violently inflammed, a vesicular eruption accompanying the inflammation and intense itching. The face was of deep orange colour, eyes closed by swelling. The inflammation extended to the edge of, but not to, the scalp." Fearful ordeal this! The poor sufferer went through his agonies "confined to the house for a week" (it might have been longer) suffering from intolerable itching. In ten days he recovered, the skin desquamating from the affected parts. The nuts were afterwards identified as Cashew-nut. "The affection corresponded exactly with the Rhus poisoning" says Dr. Prentiss, M.D., Washington, who records the case together with two other cases of poisoning from the congenor of the Cashew-nut, *R. vernicifera* and *R. toxicoldendron.* Be it noted to the discredit of Homoeopathy that one of the cases of poisoning from *R. vernicifera* occurred as the result of an administration of a Homoeopathic pellet labelled "Rhus." The other occurred from the Japanese lacquer, the sap of *R. toxicodendron.* Vesicular eruption accompanied with swelling is the prevailing characteristic of such irritation. The learned Professor wrongly says in this paper that there are two species of the Cashew-nut — *Anacardium occidentale* and *A. orientale.* My readers are already aware that *A. orientale* is not a species of the Cashew nut plant, but a distinct genus described by me as *Semecarpus anacardium* in a former volume of this Journal.

* The Therapeutic Gazette, p. 448, vol. v, 3rd series, Whole Series, volume xiii, 1889.

† There is seldom such delay in the development of the consequences, for, though the vesicating element is evanescent, the effect is quick generally speaking. There may be skins however, which are slow in being acted upon, as there are brains slow in "taking in an idea."

Reprinted from
Journal Bombay Natural History Society
Vol. 10 : 260-279. 1895

PART XIV.

ALANGIUM LAMARCKII — (*Thwaites*).

Natural Order — CORNACEAE.

MARATHI — अंकोल (*Ankol*).

This is a very handsome tree, and grows very well in the Konkan. Whether is foliage, flower, or fruit, — in whatever condition or season it is seen, — it is a striking plant. It is beautifully green-leaved throughout the year, except when about to blossom. Every branch of it, from the largest to the smallest, is covered from head to foot with clusters of cream-white, sweet-scented flowers. The entire plant thus in full bloom is very attractive to the eye. Later on, when in the height of summer the branches are loaded with clusters of fruit of the size of an ordinary marble, their rich bright crimson is particularly charming to the eye.

TRUNK. — The trunk of the tree is short, erect, generally from 2 to 3 feet in girth. Rheede gives its girth as 6 feet (Hort. Ma., vol. iv., p. 55). Sometimes the trunk is irregular. There is a large tree in the compound of the Military Hospital, Thana, the girth of which is fully nine feet. The main trunk sends out underground stems or "*suckers*". The tree in the Millitary Hospital compound referred to has nearly half a dozen distinct trees developed from such suckers within an area of twenty feet around. In a Mahomedan grave-yard not far from the Civil Hospital, Thana, there are several large trees from which several smaller trees have developed through suckers. Writing about Central India Plants, Brandis observes that the trunk "coppices well" (Forest Flora, p. 250). So it does in the Konkan.

The height of the tree varies from 30 to 40 feet; under favourable circumstances it is sometimes as much as 50 feet.

BRANCHES.— The branches and branchlets are said to be often spinescent. The Thana plants appear to be singularly free from spines of any sort. The branches are very irregular.

BARK. — The thickness of the bark as given by Brandis is only half an inch. I think I can safely say that the bark is not unfrequently about an inch in thickness. In young branches it is less than a quarter of an inch. When Brandis says that the bark is "grey with some white specks," it is to be understood that he refers to the bark of young and tender branchlets. The bark of the new off-shoots is light green. Whether in the young or old branches and branchlets, the bark is marked with irregular undulations. It is also deeply marked with the scars of the fallen leaves and pedicels of the former season.

WOOD. — The wood is described by Brandis as "light brown or yellowish-brown." "It is tough and strong, weighing 49 lbs. per cubic foot." Brandis very happily describes it further as "fine, even, close, and smooth grained." Kurz says that the wood is "dark-brown."* This answers the description of the wood examined in the Thana plants. "The wood is beautiful" says Roxburgh — (Flora Indica, p. 404, — Calcutta, 1874).

LEAVES. — Exstipulate, petiolate, alternate; membranous, says Brandis; 3-6 inches long, 1-2 inches broad; entire. Both Hooker and Brandis say they are persistent, but of this more hereafter under the head of "Remarks." The shape of the leaves is very variable even on one and the same branch — from linear oblong to elliptic, obtuse, acute, or long-acuminate; pubescent or tomentose when young; glabrous or pubescent below when full grown; the base unequal, often somewhat equal. The chief character of most leaves is that they arex three-nerved at the base. The nerves are pubescent and distinctly white and prominent on the under surface. The main lateral nerves vary from five to eight on either side of the midrib, joined by prominent transverse and intra-marginal veins. Brandis observes that in the axils of primary nerves there are often "tufts of hairs or hollow glands." This is a fit subject for microscopical workers who may be interested in the study of the morphology of the leaf of this plant. It may be observed here that in the axils of the leaves of one season

*Kurz - Forest Flora of British Burma, Vol. I, p. 543, 1877, Calcutta

there are buds of the blossom and foliage of the next season. These buds are also well worthy of a more extended microscopical study.

PETIOLE. — Brandis very accurately describes it as "1/4 inch, hairy, often villous or woolly;" not unoften it is slightly twisted on its axis.

FLOWERS. — White; fragrant; hermaphrodite; silky white, jointed on the pedicel; in axillary, close small fascicles or condensed cymes. The flowers are seldom, hardly ever, more than three at a time in each axil. PEDICELS pointed, very short; 1/8 inch.

BRACTS. — Clarke says (in Hooker's Flora of British India, vol. ii, pp. 741, 742) that the bracts are entirely absent. Brandis, on the other hand, says that "the pedicels are bracteate" (*op. cit.*). Brandis further remarks that the peduncles and pedicels are usually woolly.

AESTIVATION — "Twisted" say Wight and Arnott (Prodromus p. 325); Valvate says Clarke. This is notably so in the corolla.

CALYX. — Tubular; calyx tube woolly; minutely 5-10 toothed; "turbinate" says Brandis. Adnate to ovary; green; accrescent. The green colour of the calyx remains to the last, even in mature fruit.

COROLLA. — Very showy; greenish when unexpanded on the dorsal surface. Varying in colour when fully expanded from pale-white to cream-white, with an occasional dash of light crimson or pink.

PETALS. — Strap-shaped or linear oblong; reflexed; deciduous; varying in number from 5 or 6 to 10; this has given rise to the species described as *A. hexapetalum* and *A. decapetalum*. Wight and Arnott observe that the number of petals corresponds to the number of the segments of the calyx. "Blunt, tawny-velvetty," says Kurz.

ANDROECIUM —

STAMENS. — Deciduous; exserted; varying in length from $\frac{1}{2}$ to $1\frac{1}{8}$ inch, woolly without (Brandis). Twice as many as the petals; thrice says Hooker, and four times as much say Wight and Arnott, as also Kurz (*op. cit.*).

FILAMENTS — Distinct but short; "with long stiff hairs at base" (Brandis); "densely hirsute" says Kurz.

ANTHERS. — Very long; basified (Brandis); Wight and Arnott says "they are introrse, 2-celled, often sterile" (Prodromus, p. 325).

GYNOECIUM —

OVARY. — Inferior, globose, 1-celled, adherent to the calyx-tube firmly, surmounted by a disk. Wight and Arnott say that "the ovary is 1-2 celled." But this is not accurate. For, observe the following remark, quoted from Lindley, in Wight's Illustrations of Indian Botany, (vol. i, p. 211, Madras, 1840), with reference to the "*Affinities*" of the Natural Order *Combretaceae* — "The solitary carpel of which the fruit consists is peculiar to these" — meaning the *Combretaceae* (K.R.K.) — "and to the *Alangieae*, and neatly distinguishes these two "orders from all others of the myrtal alliance."

OVULE. — Solitary, pendulous (Wight and Arnott).

STYLE. — One; very long or "elongate" as Brandis terms it. Glabrous; uniformly cylindric; subulate, — that is to say, expanded at the base into a thick coloured disk which is fleshy and covers the top of the ovary.

STIGMA. — Large, capitate, or subglobose. Wight and Arnott simply say it is "*dilated*," but this does not convey an accurate idea. I think the stigma may be termed "many-headed," as will be amply seen from the flowering branch in my illustration.

FRUIT. — Some call it a berry; others call it a drupe, varying from 1/2 to 1 inch in vertical diameter; ellipsoid; tomentose says Brandis; crowned by the somewhat enlarged calyx-limb; quite smooth, with slight vertical ribs when dry (Thwaites). Clarke says it is closely pubescent or finally glabrous. The colour of the fruit is black says Clarke; but as will be observed from my illustration, it is rich crimson in its entire "epicarp." The "epicarp" is tough. The "mesocarp" or "sarcocarp" is pulpy and mucilaginous. The "endocarp" is bony, and separated from the sarcocarp like a putamen (Wight and Arnott).

SEED. — Oblong, solitary, pendulous.

ALBUMEN. — Ruminated; "fleshy" say Wight and Arnott.

EMBRYO. — Straight, inverse (Roxburgh).

RADICLE. — Superior, elongated.

COTYLEDONS. - Leafy; crumpled; note that Wight and Arnott say that they are "flat," and not crumpled. Note also that it is the crumpled nature of the cotyledons of *Alangium* which distinguishes it from its congener, *Marlea*, which latter has a flat cotyledon.

REMARKS

The following are the synonyms given in the Index Kewensis* —

1. *Alangium acuminatum*, Wight = *Lamarckii.*
2. *Alangium decapetalum*, Lam. Encyc. I, 174 = *Lamarchii.*
3. *Alangium glandulosum*, Thw. Enum. Pl. Zeyl., 133 = *Lamarckii.*
4. *Alangium hexapetalum*, Lam. Encyc. I. 174 = *Lamarckii.*
5. *Alangium latifolium*, Miq. ex C.B. Clarke in Hook. f. Flora Br. Ind. II, 741 = *Lamarckii.*
6. *Alangium octopetalum*, Blancho, Fl. Filip. Ed. II, 310 = *Lamarckii.*
7. *Alangium sundanum*, Mig. Fl. Ind. Bot. I, 1, 774 = *Lamarckii.*
8. *Alangium tomentosum*, Lam. Encyc. I, 174 = *Lamarckii.*

This will be considered a formidable array of synonyms. But such of my readers as have hitherto followed my previous description and will note carefully the observations embodied hereafter under this head will be able to understand that, barring all seasonal and climatic changes, which are liable to vary at all times, from year to year, the principal characters of the species above depicted under their respective synonyms have been more or less alluded to in the foregoing description of the plant under notice. The terms *hexapetalum, octopetalum, decapetalum* are indicative of the number of petals. The term *acuminatum* refers to the apex of the leaf. The term *tomentosum* refers to the existence of a more or less villous condition of the leaf and flower. The term *latifolium* refers to the breadth of the leaf. The term *glandulosum* refers to the existence of glandular bodies, developing under special circumstances in particular localities under the influence of climate. The term *sundanum* refers to the special peculiarities of the plant as it is seen growing in the Straits of Sunda.

In describing the bark of the tees as examined by Brandis in North and Central India, he says it is "grey with some white specks." It must be understood that this description is of the bark of young

*Part I, page 70, Oxford, 1893, Edited by B. Daydon Jackson under the direction of Sir Joseph D. Hooker, as the noble man - Charles Robert Darwin - to the earnest student of Universal Botany (K.R.K.)

branches; for, as the bark gets old, it assumes a brownish tinge. The oldest bark is distinctly brown-black. Some Botanists describe the bark as smooth. Brandis describes it more accurately when he says it has irregular undulations. Brandis says that the wood of *Alangium Lamarckii* is well suited for ornamental work. It is easily worked, and when properly polished, it displays "a beautiful glossy surface." It may be noted here that the petals and stamens are distinctly deciduous. This fact is not specially mentioned by previous observers, except by Wight in the letter-press accompanying his plate No. 194/1005.* This deciduous nature of the petals and stamens is well illustrated in Wight's fig. 2 in the plate just referred to. It is also shown in my illustration (plate P) accompanying this description, where, on the part of an old flowering spring marked *B*, there are two unopened buds, and below them are the persistent green accrescent calyx and the white style capped with a multilobate stigma of the third flower, from which the petals and stamens have fallen in due course. *Apropos* of this compare the following description of Baillon :- "Style girt at "base with epigynous cupular or pulvinate disk, at stigmatose apex "clavate or capitate, oftener minutely 4 - - lobate" (The Natural History of Plants, vol. VI, p. 286). Observe, as against this description of the stigma, the remark of Dr. Trimen that the stigma is large and only 4-lobed (Hand Book of the Flora of Ceylon : Part II, 1894, pp. 285-286).

The remarks which Wight and Arnott make at the conclusion of the description of the *Natural Order*, which they term *Alangieae*, are specially worthy of the consideration of those who would engage themselves in the minute study of the very interesting flower of the plant I am describing. "The portion of the torus," say Wight and Arnott, "between the calyx and ovary, to which the stamens and petals are attached, is of a different colour and texture from the above-mentioned epigynous disk which induces us to refer the latter to the style, not to the torus."**

I may add one word more with regard to the entire blossom-process in *Alangium Lamarckii*. The following is the order every year. The entire foliage of the previous year falls in the hot weather. Then

* Icones Plantarum Indiae Orientalis vol. I, 1840, Madras
**Wight and Arnott's Prodromus, p. 325

comes the beautiful sweet smelling blossom covering the tree, denuded of its leaves, from head to foot. As the fruit next forms, first green, then gradually turning beautiful crimson, capped with the uppermost portions of the toothed calyx, new foliage appears of bright light green colour before the monsoon sets in. The plant is at this time very charming to the eye. When the fruit matures, the village-boys gather round the tree in search of the sweet pulp covering the seed. It is to them - poor half-starved creatures — an out-of door repast of much relish judging from the avidity with which they gather the fruit as fallen under the tree, or picked, by an insidious ascent on the snarled branches of the coveted plant. Note that Baillon observes that "the branches of inflorescence are elongate or sometimes more or less contracted; pedicels generally contracted."*

With regard to the observations of Loudon, who describes the plant as *Alangium decapetalum*, I have this to say : Loudon says** that the plant has ten petals; the branches are spiny. His figure, however, is only a solitary flower. This is as incomplete as it is misleading. Baillon distinctly says that *Alangium* of Lamarck is *unarmed, sometimess spinescent*. The former, but not the latter is my experience in the Thana plants. Loudon observes that the plant grows in light sandy soil. Be it so. I can add that in the *moorum* soil of Thana the plant grows very well indeed. Perhaps it may be said that in moorum soil, or any similar soil, the plant thrives superbly. "Cuttings," says Loudon, "root in sand under a hand glass in moist heat." This is quoted from Loudon for the information of those who would grow the plant in a soil which has no particle of *moorum* in it. Loudon describes it as an "Evergreen." It may be so in the country from which he writes the description. On this side of India it distinctly sheds its leaves entirely about the time of blossoming. Loudon classes the plant under Myrtaceae. This was right enough according to his lights. To day we classify the plant under the natural order *Cornaceae*. There is a sufficient justification for this change according to our lights. Loudon describes the colour of the flowers of *A. decapetalum*, as "*pale-pu*," which I presume means "pale purple." If he had said it was "*Pale-ru*" instead of

* Natural History of Plants, vol. vi, page 286

** Encyclopaedia p. 468, article 1063.

"*Pale-pu*", I should have supposed that the "pale-ru" was an abbreviated form of the term "*pale-rufous*," which would have been nearer the truth; for I have often observed a reddish or pinkish tinge on the dorsal surface of some of the petals of the flowers just before expanding. This pink tinge is most marked in the petals facing the early morning sun. According to Loudon the plant appears to have been introduced into England so far back as 1779. In concluding my observations on Loudon's description of the plant described by him as *A. decapetalum*, I may add that he considers this plant to be capable of "propagation by cuttings." On this side of India we have no such contingency. The prevailling characteristic of the plant in Thana is that it throws out "suckers." In the vicinity of the plant that is now superbly growing in the compound of the Military Hospital in Thana, there are this day more than half a dozen plants — I should call them trees — which are distinctly the product of the main plant generated by means of "suckers."

It may be observed that the plant described by the elder DeCandolle* as *A. tomentosum* (Lamarck) is more like the Thana plant than any other I have yet seen described, especially as regards the characteristics of the fruit. Witness DeCandolle's own words :

"*Floribus* * * * ; *ramis inermibus* (*i.e.*, having branches without spines), *junioribus petiolis nervibusque velutinis foliis oblongis, obtuse acuminatis, subtus vernulis reticulatis. Bacca vel drupa pubescens, cortice pubescence*" (true - K.R.K.) "*livide purpureo*" (true - K.R.K.)

In describing *Alangium decapetalum* (Lam.), Sprengel says it is spinescent. *Alangium hexapetalum* (Lam.) says he, is spineless; and *Alangium tomentosum* (Lam.) is "*subinerme*" (Slightly spineless?), whatever that may mean. All these, says Sprengel, are found in the East Indies. Trimen says that *A. Lamarckii* is occasionally armed with sharp short spinous branchlets.

As observed by Baillon, with regard to the ruminate nature of the albumen, I may add that Dr. Trimen of Ceylon also observes, in describing the seed, that embryo is "straight in the axis of slightly *ruminate* albumen," and that the cotyledons are foliaceous***. It may

* Prodromus : DeCandolle : Vol. III, p. 203

** Systema Vegetabilinum, vol. II, Gottingen, 1825 (classified rightly under "Polyandria-monogynia) – K.R.K.

*** Trimen's Hand-book of the Flora of Ceylon, Part II, 1894, pp. 28-86, N.O. Cornaceae.

be observed that Dr. Trimen gives June and July as the flowering time of *Alangium Lamarckii* in Ceylon; that its bark is bitter; its heartwood hard and close grained and dark yellow. I have found that in Baillon's illustration, figs. Nos. 247-248 of *A. decapetalum* (at page 272 of vol. VI of his Natural History of Plants) the fruit is marked ribbed. Dr. Trimen distinctly notes that the fruit is not ribbed. This will be amply apparent from my figures of the fruit as copied fresh in the natural condition. Dr. Trimen describes the fruit as "purplish-red." It is a rich "lake" colour as will be seen from the copy of the Thana fruit.

The description of the fruit and seed as given by Baillon (*op. cit.*) is worthy of reproduction here, and may be usefully read in connection with my remarks embodied in the main description of the plant I have already given. It is as follows —

"Fruit, drupaceous, crowned with calyx or its scar; exocarp thin or thick flesh; putamen, more or less hard, sometimes crustaceous, 1-2 spermous. Seed oblong; integument thin; albumen fleshy, externally smooth or sometimes sinuate or ruminate; cotyledons of axile embryo foliaceous, digitinerved at base, or flat, or slightly corrugate, or sometimes contortuplicate; radicle terte superior." Let it be noted further that Baillon observes that in the *Alangium* series, though in some flowers the introrse anther dehisces by two longitudinal clefts, these clefts may be looked for right at its margin, i.e., externally instead of in the median line of the anther-cells.* The following observation of Baillon as regards the Ovary may be also usefully quoted here for the help of those who would engage themselves in the work of extended microscopical research — "The Ovary set in the cavity of the receptacle, and consequently inferior, is unilocular in the true, *Alangiums*, and encloses, inserted a little below the summit, a descending anatropous ovule with micropyle primarily superior and exterior, later lateral, afterwards slightly contorted." To this Baillon adds a very important footnote indicating that the ovule has a double envelope. The lucidity and accuracy of this description are my sole apology for such an extended quotation from Baillon. I leave it to the mincroscopic worker to judge of its utility.

*Baillon's Natural History of Plants; p. 272, vol. VI

I may observe, in concluding this notice of *Alangium Lamarckii*, that Baillon classes the *Alangium* series under the natural order "*Combretaceae.*" The reader is referred to Baillon's Natural History of Plants itself for his reasons for setting forth such an arrangement. I prefer to accept the classification of the plant under the "*Cornaceae.*"

The plant described as *Pseudalangium* by the venerable Baron Sir Ferdinand Von Mueller under what he terms N.O. Alangiaceae,* appears to be quite a different plant from the one KI am describing here. It needs but a passing notice to show that our *Alangium Lamarckii* is not to be found in Australia — not certainly in the colony of Victoria, of which the venerable Baron has been the sole botanical guide for nearly half a century.

The plant prevails on the Coromandel Coast** as will be seen from the description given of it by Roxburgh.

Turning now to the consideration of the remarks of LeMaout and DeCaisne***, I may observe that their observation to the following effect is literally correct, namely, that the woody stem is sometimes subterranean, emitting herbaceous branches. My foregone remarks regarding the main tree now standing in the Millitary Hospital Compound in Thana and many others undescribed by me but existing in the close vicinity of my Military Hospital amply bear me out in my own description of the "suckers," as also in the quotation I now cite from LeMaout and DeCaisne. These joint authors rightly remarks that "the leaves of the *Corneae* are caducous or persistent." They are caducous in the sense that they fall when it is time for the flowers and new foliage to appear. The branches at this time are bare; the leaves fall just before the blossom appears. When the blossom appears there is not a single old leaf on the tree. This is what I have already stated, and I may repeat here to emphasize the chief characteristic of the plant at the time of its renewed foliage. This renewed foliage is the striking characteristic of the plant as the fruit is maturing.

In a small *brochure* published at Mangalore in 1891, by the Basel Mission Book and Tract Depository, entitled "Five Hundred Indian Plants; their use in Medicine and Arts," at page 68, *Alangium*

* Baron Sir Ferdinand Von Mueller's Fragmenta Phytographiae Australiae, vol. II, Melbourne, 1860-1861.

** Pl. Coromand, vol. III, p. 79, plate 283

*** P. 475 of Mrs. Hooker's Translation.

hexapetalum (Lam.), has for its synonym *Alangium karangolam* (Adans.) classed under family *Alangiaceae*. At page 71 of the same *brochure* there is the following remark, which is worth reproducing — "The sage—leaved *Alangium* described by Linnaeus is also considered the *Alangium decapetalum* of Lamarck and the *Alangium acuminatum* of Wight."

In a small book, which now seems to be almost forgotten, but nevertheless valuable, published in Bombay so far back as 1840 (2nd Edition), by James Chesson in his "Times Press", under the title of "Manual of Gardening in Western India," and written by Mr. R. Riddell, there is a very important note (at page 71), which I may well reproduce here, fully believing that Mr. Riddell was then nothing his personal experience. "*Alangium decapitelum** (*sic*), native *ankool*," says he, "is a small tree with whitish flowers; *the petals vary on the same tree from six, eight to ten.* The fruit is astringent, but eaten by natives." I quote these remarks with a view to group the *Alangium* species, or call them mere varieities, if you like, under one head, *viz., Alangium Lamarckii,* no matter what the floral envelopes be, as regards the number of their respective parts, on the nature of their development. Dr. Balfour notes, what I have not seen noted elsewhere, that the wood of *A. decapetalum* is said to be peculiarly sonorous.

"In Ganjam," says he, "the leading bullock has a bell of it termed *lodoke* round its neck, the sound being heard to a great distance in the jungle."

POISONOUS PROPERTIES

Baillon remarks, on the authority of Lamarck and DeCandolle, that "*Alangium decapetalum* and *Alangium hexapetalum* are said to be purgative and diuretic."[†] Brandis says that the root is aromatic. As will be seen from my remarks later on, I have sufficient reason to believe that the bark of the root is poisonous. Only remember, pray, that I use the word "poisonous" in its widest sense. It has distinctly dangerous emetic properties, followed by a weakened action of the heart. Here may be noted what I gather from the celebrated Rheede of sacred Botanic memory. Rheede says[‡], in sufficiently distinct terms, that "the root is *acrid* and bitter." The words used as regards odour in Rheede's work, written in Latin, are "Odor gravis."I

* The proper word is decapetalum - K.R.K.

† Baillon's Natural History of Plants, vol. vi, page 279

‡ Hortus Malabaricus, vol. iv, pp. 55 56, tab. 26

have found that some English writers say, in translating these words, that the odour is "*heavy.*" I do not know whether to an Englishman or to a Briton, I should say, the word "heavy" is capable of conveying the exact sense of the Latin term "*gravis*" when it is made to mean "heavy". Here for a moment I may crave the indulgence of my strictly Botanical reader if I venture to try and determine the meaning of the Latin word "*gravis*" as used by Rheede with reference to the odour of the root bark, or, for the matter of that, the odour of anything whatsoever. I may here state, for the information of such of my readers as do not know Latin, that the Latin adjective "*gravis*, as applied to smell or flavour, means "strong," unpleasant," or "offensive." The English rendering of, it, as used by some in the word "heavy", conveys no meaning. According to the Latin Lexicographer Andrews, to whom, since 1866, I am much under obligation for my limited knowledge of Latin, the term "*gravis*," also means *bitter*. This meaning in implied in the works of M. Terentius Varro, A Roman writer on Husbandry, who flourished in the last century before the Christian era. According to Rheede, the *taste* of the leaves is acrid, but they have no odour. It may be noted in passing that the description of both Lindley and Brandis, to the effect that the root of *Alangium Lamarckii* (be it known under any of the synonyms I have detailed above) is *aromatic*, appears to conflict with the description of it given by Rheede. No Latin lexicographer has, so far as I know, given to the Latin word "*gravis*" the English equivalent of "aromatic." Here I crave the assistance of better Latin scholars, indifferent and poor as I am in my knowledge of the Latin tongue, which I studied thrity years ago, and of which I am no better master now than I was then. In describing the root Brandis is, in my opinion, more accurate (as he always is in all his Botanical utterances) when he says that the root is aromatic — for I do not think that the odour of the root is in any way "*strong*" or *unpleasant*" — I can positively say it is not "*offensive.*"

With regard to the action of the root on the alimentary canal, Rheede distinctly says it is cathartic. It produces, says he, serious and cathartic discharges from the intestinal canal.* If such is the

*In the original Latin text of Rheede the term *alvus*" is used, which I think means not only "*the belly*" or "*abdomen*", but also the '*'stomach and entrails*" "*Astringere alvum*" (Celsus I, 3) means to make costive, *i.e.*, to bind the entrails - produce constipation - K.R.K.

experience of Rheede, and of those indigenous learned writers who gatherd for him their ancient knowledge as found prevalent on the Malabar Coast, from whence he wrote in days gone-by, I may say that I cannot but congratulate myself, at the present day, in my own experience as regards the poisonous nature of the root bark, *nay even the entire root*, on this side of Western India. Rheede wrote in his day, strengthened by the researches of his native co-workers. I am writing in my day with the help of past experience, but without the special help of any co-worker. I wish I had the help to-day that Rheede commanded when he worked on the Malabar Coast. All the more do I express this wish, for I feel that there will be some who will question my taste in including this plant under the "poisonous head." But I fear no contradiction, as I crave for more co-operation in determining the poisonous nature of the root-bark of *Alangium Lamarckii*; and I wish to specially point out a dangerous property in the root bark which has not been yet experienced or specifically recognized by the recognized writers on Indian Toxicology who have preceded me. Rheede notes that the fruits are seldom eaten. "For," says he, "they heat the blood exceedingly." "Heating the blood" is a popular expression in India, and as I fear Rheede was only copying an expression of those natives of Malabar who helped him in his botanical researches, he has fallen into a popular error which is easily pardonable. Nevertheless, such an error is misleading to a student of Pharmacology and Physiology trained in an English or European school in the nineteenth century. I am not yet able to understand what the term "heating the blood" means. Perhaps it is my ignorance, and Rheede in his day knew better. All I can say is that, without the slightest fear of heating their blood, the Thana boys devour the fruit greedily. It is a distinct seasonal treat to them, judging from the avidity with which they devour the ripe fruit. Rheede's native reporters might have represented to him, according to their lights, possibly dim, that their native brethren "seldom ate" the fruit in Malabar. Possibly the tastes of their brethren in Malabar differed from those of my co-inhabitants of Thana. But that does not go to prove that the fruit of *Alangium Lamarckii should be declared* absolutely inedible. In support of *my view regarding* the edible nature of the fruit, I may quote DeCandolle. He says pointedly that the fruits of the entire *N.O.*

Alangiaceae are edible. Such is my experience; or rather my recent experiences supports DeCandolle's earlier observation. Lindley* observes that the Malays attributes purgative and hydrogogue properties to *Alagium decapetalum* and *Alangium hexapetalum*.

One word more with regard to the edible and non-poisonous nature of the fruit. I can cite two very eminent Botanists who hold the view that the fruit is absolutely free from any poisonous or "blood-heating" properties so called. Royle repeats the opinion of Lindley that the fruit is edible. Dr. Wight also says that the fruit of the *Alangiums* "is eatable, but not palatable, being mucilaginous and insipid." I may add that Brandis and others state that the fruit has a somewhat sweet and astringent taste. Such is my own experience. "The nucleus," says Rheede, "is *bitter-sweet*;" I can say that it is so.

It is well worth noticing here that instances of fruits of plants being edible, and on the other hand their roots being emetic, are not unknown. I have distinctly referred to one marked instance given by Mr. W. Bartlett† in one of my former papers in connection with this series (*vide* my paper on *Moringa pterygosperma*, in Vol. ix, p. 168, of this Journal). Mr. Bartlett refers therein to the dangerous sickness caused by eating the root of the French Bean plant.

In detailing the properties of *Alangium decapetalum* (Lam.), classed under *N.O. Alangiaceae*, Colonel H. Drury observes‡ that "the juice of the root is reckoned anthelmintic and purgative. It is also employed in dropsical cases, and pulverized, is a reputed antidote in snake-bites." This is a quotation from Roxburgh; but niether Roxburgh nor Drury mentions specifically whether the root is an antidote to the poison of the "colubrine" or of the "viperine" snakes. Symptoms of poisoning vary in each case, as is well known to those who have devoted special attention to this subject. For years past the term "snake-bite" has been very vaguely used, and passes muster in the eyes of those who would pose as discoverers of a cure for the deadly cobra bite, *i.e.*, the bite of the genuine *Naja trepudians*.

*Treasury of Botany, vol. 1, p. 720

† Pharmaceutical Journal, vol. ii, p. 721, 1st series

‡ The useful plants of India : London, 1873, 2nd Ed., p. 24 [Observe that in this work *A. tomentosum* (DeC). and *A. hexapetalum* (Roxb.) are cited as synonyms.] - K.R.K.

Surgeon-General E. Balfour of Madras also states* that the aromatic roots of *A. decapetalum* [synonyms *A. hexapetalum* (Roxb.) and *A. tomentosum* (Lam., D.C.)] are used by natives in snake-bites. This observation has the same vagueness to which I have just referred above. Surgeon Lee of Mangalore† distinctly says, however, that powder of the bark (40 grains made into a bolus) is given in cases of cobra-poisoning. He is not decided, however, in saying that it is a cure. He remarks that the root is well worth trying in cases of cobra- bite.

The following observations from the writings of Hindu writers may be usefully read in connection with the properties of the plant under notice. According to the quotation from Shiva Datta given by Kata Bhat of Junagadh (*vide* his compilation entitled "Nighant Sangraha," p. 123), the spine bearing, white-flowered, red-fruited *species*, or call it *variety*, if you will, has a thick root, which is distinctly cathartic and emetic. In Narhar Pandit's Raj--Nighanta, it is mentioned that the oil expressed from the seed also is cathartic (*vide*, p. 84, Benares Edition, 1883). In Madan Pal's "Nighanta" (Calcutta Edition of Shri Bhuvan Chandra Basu, 1886, p. 13), the plant is supposed to possess hypercathartic properties. In the yet more recent work known to Marathi readers as "Nighanta Ratnakar,"‡ it is stated in five Sanskrit *Shlokas* (stanzas) that the juice of the entire plant is emetic and highly purgative, *i.e*, productive of watery, alvine discharges. The authorship of the shlokas is not declared, but I think we may safely believe that they convey the experience of the ancients. In passing, I may observe that in this work the juice of the plant is credited with the property of curing the poisonous bites of bad snakes, and also those of "dogs, mice, and cats!" Rather is large order this, and of doubtful curative powers. The writer goes further, and credits the juice of the plant with the still more doubtful property of driving the devil (Sanskrit — *Pishach-pida*) out of human kind! I know no individual of the vegetable kingdom yet that can be said to possess such a quality. The "devil" may be safely said to have it all his own way, in spite of the curative resources of the vegetable kingdom. The "devil's kingdom" is unassailable the members of the vegetable world.

The "Encyclopaedia" of India, vol. i, p. 63; 3rd ed., London

† See Watt's Dictionary of the Economic Products of India, p. 154, vol. i, 1889, Calcutta.

‡ Published in Bombay in three volumes in 1867 by Vishnu Vasudev Godbole. (see pp. 20,21, vol. i.)

The cathartic properties of this plant are also referred to in "Dhanwantari Nighanta" (*Vide* p. 60, *Shloka* 250, advanced proofs, Edition Anandashram Series of Mr. M.C. Apte). So far as I know, Bhav-Misra does not allude to this plant in his celebrated work entitled "Bhav-Prakash."

Coming to the later indigenous writers of our own generation we have the following information. Jaikisson Indraji maintains distinctly that in large doses the root-bark is emetic, but he adds that it is safe. In this I fear he follows Mooideen Sheriff, and has no special clinical experience of his own to guide him. Dr. Sakharam Arjun does not refer to any of the cathartic or emetic properties of the plant. When he dismisses his note on this plant in his Catalogue of the Bombay Drugs in the brief manner he does, one can gather he had no personal knowledge of the use of the plant. It certainly does not grow in Bombay; and when in 1882, he saw it with me in the Thana Military Hospital, some years after he had published his Catalogue, he admired the tree immensely — the true lover of beautiful plants he ever was. A younger indigenous writer of the present day is Dr. Virji Zina Raval, I.M. and S., of the Bombay University. He notes in the his Gujarati work named "Arya Aushadha" (p. 169, 1889, Ahmedabad) that the root is diaphoretic and emetic; for the former purpose smaller doses suffice (1 to 2, *Wals*); for the latter 1/4 tola is required.* "As a purgative," says he, "the dose is up to one wal."

To Honorary Surgeon Mooideen Sheriff of Madras is due the credit of having brought to the notice of the profession of our day the emetic properties of the root-bark of *Alangium Lamarckii.* The first reference to this experience of Mooideen Sheriff appears to have been made by Dymock so far back as 1879† in the Pharmaceutical Journal of London. Watt has subsequently referred to the same in his "Dictionary of the Economic Products of India.‡ The emetic properties referred to are mentioned by Mooideen Sherif in his supplement to the"Phar macopocia of India." He says that "*the root bark has proved itself an efficient & safe emetic in doses of fifty grains; in small doses it is*

*A tola is 48 wals; a wal is about $4\frac{1}{2}$ or 5 grains.

† Vol. ix, 3rd Series, 1878-79, p. 1017 (The same reference is repeated by Dymock in the Pharmacographia Indica, vide vol. ii, pp. 165-166 - K.R.K.)

‡ Vide vol. i, pp. 154-155.

nauseant and febrifuge" (the italics are mine. — K.R.K.). The following quotations, marked (a) (b) and (c) from Watt's "Economic Dictionary" (pages quoted), may be usefully read in connection with Mooideen Sheriff's remark just quoted by me in italics —

(a) "In an official correspondence forwarded by Mooideen Sheriff to the supreme Government regarding the Pharmacopoeia of India, Dr. Mooideen Sheriff says further of this drug : "It possesses the emetic and nauseant properties of ipecacuanha." (b) "It is a good substitute for ipecacuanha, and proves useful *in all the diseases** in which the latter is indicated except dysentery." (c) "Doses as an emetic from 45 to 50 grains; as a nauseant, diaphoretic and febrifuge from 6 to 10 grains; and as an alterative tonic from 2 to 5 grains." These points will be referred to later on. There is one observation in Watt,s citations in his Dictionary from writers who have helped him in the compilation of his cyclopaediac work, which I cannot pass unnoticed : "Surgeon Joseph Parker, M.D., Poona," as he then was, is a well known brother-officer of the Indian Medical Service, now a Surgeon-Lieutenant-Colonel, and the Medical Store-keeper of Bombay. On the testimony of this officer, not sanctioned by personal experience, as the officer himself candidly says, Watt enters in his dictionary a remark which I think is of doubtful accuracy. It is to the following effect — "The oil of the root-bark is said to be a useful external application in rheumatism." Should this happen to meet the eye of Surgeon-Lieutenant-Colonel Parker, I should like to know if he is still of the same opinion; and I should be thankful to know if I am wrong in maintaining the view that, botanically speaking, there is no source for any oil in the structure of either the root or the root-bark. So far as I can ascertain at present, I might say that one might as well "by repeated efforts extract oil from the sea sand" (as the Marathi proverb goes) as expect to get any oil from either the root or root-bark of *Alangium*.

With reference to the quotations (a), (b) and (c) cited above from Mooideen Sheriff, I may dispose of his remarks under (b) and (c) by saying that they refer to points which I cannot very well discuss in this journal by unnecessarily occupying its pages on purely medical

*The italics are mine – K.R.K.

and clinical questions. Other journals of a less popular character are open to me for a discussion, which could hardly be expected to be of any importance or use to the readers of this journal. I wish to direct my remarks in the following pages to what I consider the Kernel of my observations under the head of *poisonous* properties in the description of this plant.

The chief point, then, that I wish to impress upon the readers of this journal is that, recognizing, as I do fully, the emetic properties of the root-bark, I have sufficient reasons, from personal experience, to be able to say that, although the root- bark may, and does in my experience, prove an efficient substitute for ipecacuanha *as an emetic*, I cannot endorse the opinion of Mooideen Sheriff "*that it is safe in doses of even fifty grains.* I would rather use ipecacuanha as an emetic-certainly not *Alangium* root instead. The officinal dose of ipecacuanha powder as an emetic is 15 to 30 grains. Dr. A.B. Garrod* distinctly says that " in large doses it has considerable power in lowering the circulation." This veteran therapeutist includes the drug amongst those that act as *sedatives on the vessels and capillary system.* Dr. Lauder Brunton says that "ipecacuanha is one of the drugs which, while it increases the secretion of the lungs and bronchi generally, tends to *Lessen the blood-pressure.*" In the earlier portion of his work (p. 219) he says[†] with regard to *Emetin*, which is the uncrystallized active principle obtained from ipecacuanha, that; "when injected either subcutaneously or into the veins, it produces death by *cardiac paralysis.* It paralyses the vessels first and then the heart, so that the blood-pressure sinks to zero, while each cardiac pulsation is still powerful." Dr. Ringer[‡] says "that in large doses it produces both nausea and vomiting, and, like other emetics, some *general weakness with sweating.*" On the authority of M. Chouppe and Dyce-Duckworth, Dr. Ringer further says that "emetin acts through the terminations of the pneumogastric nerve," and that it "without doubt enters the blood." According to Dr. Ringer, ipecacuanha is a distinct depressant.

Sir Robert Christison, writing so far back as 1845, considers that *emeta* (which is the term uses for *emetin* of the present day) "is a

*Materia Medica, p. 396, 3rd Ed. 1868.
† Pharmacology, 1885, p. 867
‡ Therapeutics, 1880, 8th Ed., pp. 406 to 419

powerful poison. Two grains of the pure alkaloid will kill a dog; and the symptoms are frequent vomiting, followed by sopor and coma, and death in fifteen or twenty-four hours. In the dead body the lungs and stomach are found inflamed. The same effects result from injecting it into a vein, or applying it to a small wound (*Magendie*). It appears then to be a marcotico-acrid. But its irritant properties are so prominent that it might be properly arranged with the vegetable acrids."*

Dr. Alfred S. Taylor, in his work on Poisons†, makes the following remarks regarding *emetin* — "Pelletier and Magendie found that from 6 to 10 grains of the impure alkaloid given to animals caused violent vomiting, followed by stupor and death in about fifteen minutes. On inspection the alimentary canal was observed to be inflamed."

I should not have deemed it fair to occupy the attention of my readers with such elaborate quotations on the poisonous properties of *emetin* did I not think that the active principle of *Alangium Lamarckii*, from my experience of fifteen years, possesses in a pre-eminent degree the properties of *emetin*, whether as a vascular sedative or as a "narcotico-acrid" as very appropriately termed by Christison. In my experience the root-bark powder of *Alangium Lamarckii*, besides being a sure emetic, has a decidedly more powerfully sedative effect on the human heart and blood vessels. It is besides productive of violent irritation of the mucous coat of the stomach, followed by gastric catarrh lasting many days.

The recent researches of the writers of the "Pharmacographia Indica" (Dymock, Wardell and Hooper)‡ have, happily for me isolated a very bitter non-crystallizable alkaloid, which they have provisionally called *alangine*. It is to be hoped that, now that Dr. Dymock is gone, Dr. Wardell and Mr. Hooper may be able to establish the identity of *Alangine* with *Emetin*. I leave it to other future investigators also of the pharmacological properties of Indian plants to decide this point.

"The bark of Dogwood," a congener of *Alangium*, especially of *Cornus florida*, say Le Maout and DeCaisne,# "is bitter and astrin-

* A Treatise on Poisons, p. 892, 4th Ed., Edinburgh

† Ed. of 1848, p. 759

‡ Vide p. 166, vol. ii

Mrs. Hooker's Translation of their General System of Botany, 1873, page 477

gent, and yields a principle (*Corniine*), which is administered in North America instead of quinine." Note that the recent researches of Sohn term this active principle Cornin.* It is neutral and bitter, and consists of crystalline silky needles, soluble in water, alcohol, or ether. (*Ann. Chem., Ph. 14, 206 &c.) It does not appear to possess any poisonous properties.*

The next question I would suggest to future workers in pharmacology is as to whether *alangine* can be classed under the group of *Saponins*† , which in small doses cause symptoms in man "closely resembing the nauseant stage of emesis, that is, a raw feeling in the throat, tendency to cough, and increased secretion of mucus."

To sum up, the result of my experience is that, even although the root-bark of *Alangium* is a sage substitute for ipecacuanha as a mere emetic, as stated by Mooideen Sheriff, it is a more powerful cardiac sedative than ipecacuanha is reputed to be. As compared with ipecacuanha, I think *Alangium* is a tardy and not a prompt emetic. From the very fact that ipecacuanha is prompt emetic, one may consider it safe, as it does not remain in the stomach, but is discharged from it before a large quantity of it has time to be absorbed by the blood. Emesis being slow in *Alangium*, the root-bark of it has more time to be absorbed by the blood. Hence, probably, its pronounced action on the cardiac inhibitory nerves.

* Dictionary of the Active Principles of Plants : London, 1894, page 45, § 70.
† Schmiedeberg's Pharmacology (translated by Dixon, 1887), page 68.

Reprinted from
Journal Bombay Natural History Society
Vol. 10(3) : 482-502. 1895

PART XV.

CORALLOCARPUS EPIGAEA. — *Hook. fil. in Gen. Pl. I, 831.*

Natural Order — CUCURBITACEAE.

MARATHI — कडवी नाई KADAVI NAYI

This is a small herbaceous annual climbing plant throwing out foliage in the early part of the rainy season, flowering between July and September; and maturing into seed in the cold weather.

ROOT. — Very large; said to be turnip-shaped, but oftener irregular.

STEM. — Roundish; rather succulent; prostrate glabrous; glaucous (Trimen). Seldom as thick as a quill; jointed; often bending at right-angles at the joint, or "flexuose" as Wight calls it.*

TENDRILS. — Simple; in the older branches very much branched; distinctly lateral.

LEAVES. - Rather small; 1-$1\frac{1}{2}$ inches long; rotundate; very cordate, sometimes, at the base; more or less 3-5 lobed. Lobes obtuse or acute; irregularly and slightly dentate. They are "rounded," says Wight; the lateral ones the broadest and slightly 2-lobed. Both the surfaces of the leaf are rather thick or fleshy; and shortly or roughly pubescent. Wight says the leaves are "sometimes only obtusely angled, densely covered on both sides with short *bristly hairs*." This is something more than being "pubescent" as described by Trimen. Wight's description is more accurate as regards the Indian species.

PETIOLE. — More than half the length of the leaf; stout, glabrous; and somewhat round. It is often so bent as to give the leaf a deflexed appearance.

---

* Icones, Vol. II, p. 7. Pl. 503/1079, 1843

FLOWERS. — Unisexual; monoecious. Yellowish-green, appearing between June and September.

MALE FLOWERS. — Very small. Shortly racemose at the apex of a long thickish peduncle, on short pedicels; decidedly more yellow than the female flowers. They are at time solitary.

INFLORESCENCE. — A Raceme. "A small corymb at the apex of a long peduncle," says C.B. Clarke.* This is a more accurate description.

PEDUNCLE. — Roundish, smooth, straight, stiff; having the flowers nearly all crowded at the apex; length $1\frac{1}{2}$-$2\frac{1}{2}$ inches. Clarke says it is $1\frac{1}{2}$-2 inches. From the plants I have been able to examine, I may say that the length is not more than an inch at the utmost.

BRACTEOLES. — Present at the insertion of the pedicels.

PEDICELS. - Short; "$\frac{1}{4}$-$\frac{1}{2}$ inch or sometimes 1 inch" (Clarke).

CALYX. - Campanulate (Wight); lobes, 5, short.

COROLLA. — 5-partite. The calyx-lobes are "undivided," says Arnott.† But this would be considered due to insufficient development.

STAMENS.

FILAMENTS. — 5 (*arranged* 2 + 2 + 1); *i.e.*, two sets of stamens united forming a bundle, each of two filaments; and one solitary; nearly sessile; at the mouth of the Calyx-tube.

CONNECTIVE. — Very small.

ANTHERS. — Free; 1-celled. Arnott says‡ that the anther-cells are "straight oblong, lateral on their thin connective which is *not Produced.*" There is no rudimentary gynoecium in the male flower (Baillon). This may be considered as prevailing characteristic of this genus.

FEMALE FLOWERS — Axillary; solitary; in the same or different axils from those of the male flower (Wight); or "accompanied with a male raceme" (Trimen). Larger than the male; more greenish than yellow. Sometimes the female flower arises from the axil of the tendril.

PENDUCLE Short.

CALYX. & COROLLA. — as in the male.

OVARY. — Inferior; ovoid; of three carpels united into one cell.

* Hooker's Flora of British India, vol. II, p. 268

† Vide Hooker's Journal of Botany, vol. III, 1841

‡ History of Plants, Vol. VIII, P. 402.

OVULES. — From 6 to 9 in number; horizontal.

DISK. — Absent.

STIGMAS. — 3.

STYLES. — 3, distinct but very short.

FRUIT. — $\frac{3}{8}$ inch, often $\frac{1}{2}$ to $\frac{3}{4}$ inch; roundish or ovoid; concial, says Trimen*; with a blunt beak which is hardly perceptible in most female flowers. The fruit is smooth. Its colour varies from orange to brick-red. which is most marked between the two ends — *i.e.*, in the mid-portion, bearing four or five vertical mottled-whistish streaks. The top of the fruit is greenish. Botanically, the Fruit is an "Inferior Berry." But of this more hereafter, under the head of *Remarks*.

SEEDS. — Elliptical or nearly globose; little-margined, about 6 in number; much longer than broad (Clarke). Trimen says the seeds are "about 9, in an orange-coloured pulp," very slightly compressed; pale brown; "adpressed-pilose." Wight on the other hand says that the seeds are white. In the specimens I have examined, the pulp is distinctly whitish and gelatinous. The colour of the seed varies from white to brown. The older or maturer the seed the browner is the colour. Wight and Arnott in their Prodromus (Vol. I. p. 346), say the seeds have slightly "Compressed sides."

ALBUMEN. — Absent.

REMARKS.

The following is a list of the various synonyms of the plant suggested by authoritative Botanists from time to time.

1. *Corallocarpus epigaeus*, Trimen's Handbook of the Flora of Ceylon, Vol. II, p. 258, 1894.

2. *Bryonia epigaea*, Blume, Bijdr. 925 = *Methria Rumphiana.*

3. *Bryonia epigaea*, Rottl. in ges. Naturf. Freunde, Neue Schr. IV. (1803) 223.

4. *Bryonia epigaea*, W. & A. Prodromus, Vol. I, p. 346.

5. *Bryonia epigaea*, Willd., Sp. Pl. IV, 619.

6. *Bryonia glabra*, Roxb. Hort. Beng. 104, and Flora. Br. Ind. III, 725.

7. *Bryonia palmata*, Wallich's Cat., No. 6711 (D).

8. *Bryonia sinuata*, Wall. Cat., No. 6711 (D), and Bentham and Hooker's Gen., Pl. I, 831.

* Hand Book of the Flora of Ceylon, Vol. II, p. 258, 1894.

9. *Æchmandra epigaea*, Arnott. (*Vide* Hooker's Journal of Botany, Vol. III, pp. 271-274, 1841).

10. *Æchmandra epigaea*, Wight's Icones, *t.* 503.

11. *Æchmandra epigaea*, Dalzell and Gibson's Bombay Flora, p. 100.

Such is the formidable array of the synonyms.

Following Hooker *fil.*, I have named the plant I am describing *Corallocarpus epigaea* instead of the more correct modern combination *Corallocarpus epigaeus* used in the "Kew Index", and by Trimen in his recent Flora of Ceylon. Unfortunately our botanical nomenclature has not as yet reached that state of grammatical purity which should absolutely discard the combination of a generic name with a masculine termination and a specific name with a female termination, or *vice versa*. In naming the plant I am describing, I have adopted the terminology as given in our standard work on Indian Botany, namely, Hooker's flora of British India, without attempting to question its grammatic propriety; and I did so long before Trimen's Flora of Ceylon or the Kew Index had made its appearance. Dr. Dymock in his "Pharmacographis Indica" adopts it, and I have adopted it accordingly.

Dr. G.A. Walker-Arnott, in his very valuable contribution to Hooker's Journal of Botany in 1841, cited above, was the first among Indian Botanists to acknowledge Schrader's attempt to subdivide the *Cucurbitaceae*, in anything like a scientific way. Schrader's arrangement, then new, was published in the "Linnaea," XII, p. 401. The stigma in *Bryonia scabra* (var. E.M.) has the style trifid, and, says Arnott, "the stigma precisely as in *Bryonia dioica.*" I have already observed that "*the connective is small.*" Arnott observes the connective in the species belonging to the genus *Bryonia* is "sinuated and lobed," and that the anther-cells are placed on the back margin of the connective. Arnott, moreover, observes that the style in genus *Bryonia* is "surrounded at the base with a thick, annular, fleshy, usually lobed, disk." Following Hooker (C.B. Clarke's Article, cited above, in "Flor. Br. Ind."), and, judging from my own experience, I have said that the disk is absolutely absent.

Historically, the formation of the genus *Æchmandra*, of which the plant I am describing formed species under the name *Æchmandra epigaea*, was the result of the classification originally suggested by

Schrader. Whether we know the plant as belonging to the genus *Æchmandra* or to the genus *Bryonia*, one thing is certain, that *Corallocarpus epigaea* is of the same genus as the English *Bryonia dioica*. C.B. Clarke completely drops the genus *Æchmandra* from his classification of the *Cucurbitaceae* in Hooker's Flora of British India.

The fruit of the plant I am describing is ordinarily termed a "Berry," but this term I have not adopted above in my description without modification. The term *berry* is a puzzling term in botany. It is easy enough to understand what a "berry" is in popular parlance; but in botany the term *berry* is used by several eminent Botanists with such varied significations that it would be better if the term were entirely obliterated from the Scientific Botanical Vocabulary, for more than one reason. It conveys no accurate idea as to whether the fruit is *superior* or *inferior*. It gives no idea whether the fruit is pulpy inside or has a hard stony seed. It gives no idea as to whether the fruit contains one seed or more than one. To show to the general reader of botany how many descriptions can be given from various authoritative writers, of one single term, having a sort of mixed position in popular parlance and in botanic nomenclature, I crave the indulgence of the editor of this journal to embody in the following few pages, quotations from various Botanists, of eminence. I crave the same indulgence of the reader of this journal, and trust that I may not be charged with a vain-glorious desire to parade my reading in botany, but that I may be credited with a sincere desire to expose the grievance of the systematic student of scientific botany that our botanical technical terms are sometimes extremely puzzling.

Sir Joseph Hooker's definition (p. xx. "Outlines of Botany," Vol. I, 175; "Flor. Br. Ind.") of the term "Berry," is as follows : "It is a succulent (indehiscent) fruit," in which the whole substance of the pericarp is fleshy or pulpy, with the exception of the outer skin or rind, called the *epicarp*. The seeds themselves are usually immersed in the pulp; but in some berries the seeds are separated from the pulp by the walls of the cavity or cells of the ovary, which form, as it were a thin inner skin or rind called *Endocarp*. It is not mentioned here whether the fruit is *Superior* or *inferior*.

In Lindley's "Glossary of Technical Terms" (see page xii), the following is the description of a Berry — *Berry* or *Bacca*, "that is to say, a succulent seed-vessel, filled with pulp, in which the seeds nestle, as in *Solanum*." In a companion work of Lindley's entitled "Elements of Bombay" (page 107, 1847), a *Berry* is "a succulent fruit, the seeds of which lose their adhesion when ripe, and lie loose in pulp; ex., a gooseberry." There is the same omission here as in Hooker, as to whether the fruit is *superior* or *inferior*.

Professor J.H. Balfour of Edinburgh includes the Berry (bacca) among the indehiscent Syncarpous fruits, and describes it as follows — "A *Berry* or *Bacca* is the name given to all indehiscent Syncarpous fruits, the seeds of which are immersed in a pulpy or fleshy mass. Such fruits are collectively called baccate or berried. In the true berry, such as the Gooseberry or Currant, the calyx adheres to the fruit, and the placentas are parietal, while in the Grape (Uva) the ovary alone is present, and the placentas are central. Instances of baccate fruits are seen in Solanaceous plants, such as the Potato, Egg- plant, Mandrake, Belladonna, and Winter-cherry, as well as the Mistletoe." (p. 287, Class Book of Botany, 3rd Edition, 1870, Edinburgh). Note that the *superior* or *inferior* nature of the ovary is not alluded to here.

In Henfrey's "Elementary Course of Botany" (4th edition by Maxwell T. Masters and A.W. Bennett, p. 147, 1884), the following is the description of the "Bacca or Berry" — "The *Bacca*, or true berry, is an inferior succulent fruit, crowned by the withered teeth of the calyx; it is uniformly pulpy, with a thin skin, the numerous seeds being imbedded in the pulp, as in the gooseberry and currant, &c. The term *Baccate* is, however, now generally applied to all succulent fruits, whether *superior* or *inferior*, which have not a distinct stone-like drupe." Here, there is greater accuracy.

In Professor Bentley's "Manual of Botany" (4th edition, 1882, pp. 311 and 312), the term is thus described — "The Bacca or Berry is an inferior syncarpus fruit, with a fleshy indehiscent pericarp, one or more-celled, many-seeded, pulpy. The pulp is produced from the placentas, which are parietal and have the seeds at first attached to them; but these become alternately separated and lie loose in the pulp. Examples may be found in the gooseberry and currant." * * * "The name *baccate* or *berried* is applied by many Botanists to any fruit of a pulpy nature."

Remember, as observed by Bentley, that the fruit styled by him "Pepo," which is an "inferior, one-celled, many-seeded, fleshy or pulpy fruit," has seeds which are "attached to parietal placentas," and "imbedded in pulp, but *they never* become loose, as in the case of the berry."

In the Manual of Botany, just published by Professor J. Reynolds Green, F.R.S., D.Sc., F.L.S., of the Pharmaceutical Society of Great Britain (*Vol. I, June, 1895*), and based upon the Manual of the late professor Bentley, just referred to, Professor Bentley's classification seems to be considerably altered as regards terminology. The "berry" accordingly is described as one of the "*true indehiscent, monocarpellary fruits.*" Professor, Reynolds Green, in further detailing the characteristics of the fruit termed "Berry," observes as follows* — "Though the berry is usually formed from a syncarpous ovary, there is one form of it which may be included here. It is a succulent fruit with a single cell in which is a single seed. This is hard and stony, and, at first sight, may be taken to be the same thing as the stone of the plum. It is not so, however, as the latter is hollow and enclose the seed. In the stone is the seed. This form of fruit is seen in the date and some other palms." Thus, according to Professor Reynolds Green, no fruit of the Cucurbitaceous order can be classed under the term "Berry," as it is not *monocarpellary*, but *tricarpellary*. Note that in Hooker's "Flora of British India," Clarke, our Immediate Past President of the Linnaean Society, who writes the articles on the *N.O. Cucurbitaceae*, distinctly says that the fruit of *Corallocarpus epigaea*, is a "Berry". The difficulty to me is, as it doubtless would be to Mr. Clarke himself when reading this new description of Professor Reynolds Green's that the *Berry* is now classified by the latter as a *monocarpellary* fruit-meaning that it is made of *one carpel*. I do not think the term *monocarpellary* can be made to indicate a fruit which consists of *one cavity* formed by the union two more cells. In the *N.O. Cucurbitaceae*, the fruit is distinctly formed by the union of three carpels, whereas *monocarpellary* would mean formed of one carpel only. Owing to this confusion connotation botanic term "Berry," as already observed, I have refrained from distinctly classifying the fruit

* P. 244, "Manual of Botany," Vol. I; Morphology and Anatomy, 1895.

of *Corallocarpus epigoea* as merely a *Berry*, and have used the distinguishing expression "*Inferior Berry*" to denote that the fruit is formed from an inferior ovary. But I must distinctly discard the term *monocarpellary*, as suggested by Professor Reynolds Green, in applying it to the tricarpellary fruit of any member belonging to Cucurbitaceous family.

LeMaout and DeCaisne* say that the fruit of the *N.O. Cucurbitaceae* is a "fleshy berry (rarely dry)." According to these Botanists a berry is either "simple" or "compound". The "simple" berry† is found in *Berberis* and *Arum*. It differs from the "compound" berry by originating in a "solitary carpel." * * *‡ Further LeMaout and DeCaisne observe that the fruit termed "*Berry (bacca)*, whether simple or compound, is succulent, indehiscent, *and has no stone*." (*N.B.* — The italics are mine. - K.R.K.). "It differs from the capsule only in its fleshy consistence (and not in being an inferior fruit? — K.R.K.; for be it noted that the fruit of the *Cucurbitaceae* is distinctly inferior, whereas a capsule is superior. — K.R.K.). I refrain from further quoting fully, as I should like to do, the observations of LeMaout and DeCaisne for fear it would involve the ordinary reader of these pages into a fearful maze as regards the true botanical significance of the term "Berry," and would in no way help in deciding whether the fruit of the plant I am now describing is or is not a berry pure and simple in the ordinary acceptation of the term.

The best description of the fruit of the *Cucurbitaceae* is given by Gaertner. I shall refer to him later on.

According to Asa Gray the term "berry" (bacca) "comprises all simple fruits in which the pericarp is fleshy throughout. The grape, gooseberry, currant, cranberry, banana and tomato are familiar examples."§

This is what Professor Otto W. Thome says regarding the nature of the *berry*. "In the *berry* all the layers of the pericarp are fleshy and succulent, as in the grape, currant and gooseberry; or the outer

* A General System of Botany, translated by Mrs. Hooker, 1873, p. 451.

† Op. cit., p. 99

‡ Op. cit. p. 103

§ P.299, Botanical Text Book, Vol. I; Structural Botany, 6th Edition, 1879, New York layers are harder as in the citron and gourd (the latter sometimes called a *pepo*)."*

Professor Sydney H. Vines, of Oxford, gives the following description of the *berry* (bacca)†— "The endocarp is soft and juicy, as well as the mesocarp, so that the seeds are imbedded in the pericarp : there may be one seed only as in the date, or many, as in the gourd, currant and grape : the fruit may have one loculus, as in the grape and the gourd, or several loculi, as in the orange : *and further it may be* superior, *as in the grape, orange and lemon; or inferior, as in the currant, the gooseberry and the gourd.*" (The italics are mine. — K.R.K.). If this description is taken as our guide, as we should, considering that it is the best and most accurate description I have yet seen in a student's text-book, the fruit of the plant, I am now describing may be safely termed an *inferior berry.*

Professor A. Kerner von Marilaun, of the University of Vienna, in his excellent work entitled the "Natural History of Plants," translated into English by Professor F.W. Oliver, of University College, London (Vol. II, p. 427, 1895, London), describes the *berry* as follows — "When the seed-case derived from the pistil becomes altogether fleshy and succulent, the fruit is termed *berry.* From inferior pistils arise inferior berries. From superior pistils superior berries." This description may be usefully read side by side with that of Professor Vines.

Gaertner in his description of the *pepo*‡ divides the fruit into two classes—

(1) *Solidi, i.e., solid*; having a single cavity. In this class he includes "Bryonia," which is one of the former generic names of the plant I am now describing as *Corallocarpus epigoea; Cucumis, Cucurbita* and *Trichosanthe* are some of the other genera included by Gaertner under this class.

(2) *Cavi, i.e., hollow,* under this head he includes *Momoridca charantia.*

Having, so far dwelt on the difficulties of the student of botany in giving an exact name to the kind of fruit seen in the plant I am

* P. 153, "Text Book of Structural and Physiological Botany," translated by A.W. Bennett, 1877, London.

† "A Students Text Book of Botany," London, 1895, p. 532

‡ P. XCVII, Vol. I, "De Fructibus et Seminibus Plantraum" Lipsiæ, 1801.

describing, I now proceed to examine the indigenous literature on the subject. I must premise, however, that my knowledge of Sanskrit is very limited; but what I shall hereafter say regarding native writers is all the result of my personal study of such writers.

Corallocarpus epigoea does not appear to have been described by many of the older Sanskrit writers, as it is not found in "Râj-Nighant"; "Madan Pâl Nighant"; or "Bhâv-Prakash." It is not referred to in the *Ashtâng-rhidaya* of Wâgbhat, and it is doubtful if it can be found in Charak or in Sharangdhar. The only author who describes it is Shodhal. He is referred to by Kata Bhat of Junagadh.* Katâ Bhat says that Shodhal flourished after Narhar Pandit of Kashmir, the writer of the celebrated "Râj-Nighant." Shodhal's work is based on Dhanvantari-Nighant, although the year in which his work was written is not stated. Shodhal calls the plant I am describing *Nâhi.* This word is not to be found in Bothlingk's Wortbuch or in Burnouf's Sanskrit and French Dictionary. It is from the Sanskrit word *Nahi* that the Marathi designation of *Nâi* is derived; the prefix *Kadavi* (which means bitter), as given by me above, merely signifying the bitter quality of the mucilaginous or gummy exudation obtained on section of the root in its fresh condition. Let me add here that Colonel Drury says that the root has a "subacid" taste.†

I cannot refrain from observing here that I am very much puzzled how it is that Dr. Dymock and his colleagues‡ have fallen into the error of making the following remark : "This plant is called in Sanskrit Chhilihinda, Patala-garuda, and Maha-mula or Great- root." I say that it is a mistake to say so; for I am positive that all these terms, embraced in my quotation from Dr. Dymock's "Pharmacographia Indica," are to be found in "Bhav-Prakash" (p. 111, Jugadishwar Press edition, Bombay, 1891). These Sanskrit terms refer to a plant belonging not to the *N.O. Cucurbitaceae*, to which belongs the plant I am describing, but to the Natural Order *Menispermaceae.* The name of the plant implied by the Sanskrit terms of Dr. Dymock is *Cocculus villosus* — known in Marathi as *Tanicha-Vel, i.e.*, the creeper called *Tani.* In the Marathi work called "Nighanta Ratnakar"§

* P. 569, Section 479, Nighant Sangrah, 1893

† " The useful plants of India", London, 1873, p. 88

‡ Pharmacographia Indica," Vol. II, p. 90, 1891

§ P. 123, Vol. I, Bombay edition, 1867.

Chhilihinta or *Chhilihinti,* or *Chhilli* is *Tanicha-Vel.* At p. 201 of the same volume *Patal- garudi* is also called *Tanicha-Vel;* similarly at p. 269 of the same volume, *Mahamula* is a synonym for *Tanicha- Vel.* No further indigenous support is needed to establish my view that Dr. Dymock's remark is an erroneous one, especially as Shodhal's description of *Nahi* is to my mind so complete as to make me firm in the belief that the plant he calls *Nahi* is what we call in Bombay *Kadavi Nai* = the *Corallocarpus epigoea* of Hooker *fil* ("Flora Br. Ind.," Vol. II. p. 628).

I cannot venture to say how it is that Dr. Dymock and his colleagues fell into this pardonable error, but I feel bound to point it out as the result of my study regarding the plant I am describing. That Dr. Dymock has himself identified *Tan-Wel* (or as styled in Marathi Tenicha-Vel, and briefly *Tan- Vel*), *as a member of the N.O. Menispermaceae, and not Cucurbitaceae*, will be amply borne out by a reference to p. 190, last line, Vol. II, of this Society's Journal, 1887, where, as synonyms for *Cocculus villosus*, DC., Brigade. Surgeon W. Dymock, in his valuable article headed *Marathi Names of Plants (with a glossary)*, gives the following Bombay names — ", Vasanvel; Parvel; Tan." (The Marathi word should have been and not, which latter has no meaning in Marathi. - K.R.K.) For remember that is an abbreviation of Tan-Vel or Tanicha-Vel = the creeper *Tan.*[§]

Refer also in this connection to the then Hon'ble Mr. Justice Birdwood's "Catalogue of Matheran and Mahableshwar Plants."* Mr. Justice Birdwood's remarkably accurate knowledge of the Bombay Flora is my sole guarantee for quoting his unquestionably high authority in solving this somewhat ungracious question as against Dr. Dymock, my former Teacher of Materia Medica in the Grant Medical College, in the early seventies, and my subsequent friend and brother-officer in the Indian Medical Service. Mr. Justice Birdwood, under the *N.O. Menispermaceae*, gives *Cocculus villosus* as the Latin equivalent of (Tan), Vasan-Vel (Sanskrit *Vasadani*) sic. [Note that in Sanskrit it is not *Vasadani*, but *Vatsadani* or *Vasani.*[†]]

* P. 112, Vol. II, 1887, Bo. N.H. Society's Journal

† Vide "Raj Nighant" of Narhar Pandit; Burnouf's Sanskrit and French Dictionary and Boothlingk's Wortbuch.

I cannot refrain from quoting here the remarks I have come across regarding the genus *Bryonia*, under which Loudon describes the species I have called in this paper *Corallocarpus eipgaea*, as *Bryonia epigaea.** Loudon derives the generic name *Bryonic* from the Greek Bruo = to push or grow rapidly, in allusion to the manner of its growth. * * * "The root" (presumably of the whole genus) grows to a vast size." That it does grow to a large size, and in a very irregualr manner too, I can vouch for in the plant which I am now describing, and which was in the days of Loudon known as *Bryonia epigaea*. *"Gerarde says," observes Loudon, "that the Queen's Chiefe Chirurgeon, Master William Goodorous, showed me a root hereof," i.e.*, either of *Bryonia alba* or *B. dioica* — I am unable to ascertain which, "that waied halfe an hundred waighte, and of the bignesse of a childe of a yeere old." Loudon observes further as follows :- "To this Linnaeus ascribes the quickness of its growth, though it springs late. The roots have been formerly by imposters brought into an human shape, carried about the country and shown for mandrakes to the common people. The method which these people practised was to open a young thriving bryony plant, being careful not to disturb the lower fibres; to fix a mould such as is used by those who make plaster figures close to the root, fastening it with wire to keep it in its proper situation, and then to fill in the earth about the root, leaving it to grow to the shape of the mould, which is effected in one summer."

Young Indian collectors of the *mandrake root* so-called, beware ! The delusion and deception that Loudon has described is practised in India also. Human nature is the same all the world over, whether it is for cunning men to practise delusion on credulous men, or for credulous collectors of curiosities botanical to be deluded into deception of the most unsuspected yet atrocious kind. I know of one individual at least among our young Indian Botanists who has fallen an unsuspecting victim to this trick of field collectors of botanical specimens for the mere purposes of trade.

The plant I am describing is by some native writers called *Shiva-lingi*, to denote the peculiar appearance of the seed of the plant, ovoid, and marked with a sharp convexity on one of the sides. But according

* An Encyclopædia of Plants, London 1829, pp. 810 and 811.

to the old Sanskrit writers I prefer to reserve this term, originally known as *Lingini* or *Lingika* for the ground-seeking creeper bearing a much smaller fruit, in size not larger than a pea, and known in the Konkan as Kavaliche-dhole (*Bryonia laciniosa**).

Let me observe here that Dr. Dymock and his colleagues have fallen into another slight, yet unhappy, error† regarding the interpretation of the Marathi term *Shivalingi* or *Lingini.* It is as follows — They say that "The Marathi name signifies the *Linga* of Shiva, and is an allusion *to the shape of the fruit.*" Nothing of the kind. As already observed, the fruit of *Corallocarpus epigoea* is ovoid, roundish, or conical. It is the seed of *Corallocarpus epigoea* that has the shape of the *Linga* of Shiva.

POISONOUS PROPERTIES

The observations of Dr. Dymock and his colleagues, as embodied in their "Pharmacorgraphia Indica," are extremely useful in understanding the active principles contained in the tuber of *Corallocarpus epigaea.* They have practically decided the question that the bitter principle contained in this plant is *Bryonia,* which Walz discovered in the European Bryony root. Dr. Dymock and his colleagues say that they have been "able to confirm this, by finding in the decomposition products two resinoid bodies, differing in their solubility in ether. *Bryonia* is a glucoside resolved by boiling with dilute sulphuric acid into glucose and two amorphous bodies, *Bryoretin,* soluble in ether, and *Hydrobryoretin,* insoluble in ether but soluble in alcohol." The formulae they give are as follows —

$C_{48}H_{34}O_{21}$	=	$C_{21}H_{35}O_7$	+	$C_{21}H_{37}O_8$	+	$C_6H_{12}O_6$
Bryonin		*Bryoretin*		*Hydrobryoretin*		*Glucose.*

It would be useful here to show the various formulae given by other writers for Bryonin :

Dragendorff's *Bryonin* has this formula — $C_{48}H_{80}O_{18}$

Sohn's (Walz) " " " — $C_{48}H_{80}O_{19}$

* Scc Katâ Bhat's "Nighant Sanghrah," Junagadh , 1893, pp. 122 and 123; and Apte's edition of "Dhanvantari Nighant", p. 330.

† "Pharmacographia Indica," Vol. II, p. 91; Bombay, 1891.

Masson's *Bryonin* (in *Addenda* to Sohn's "Dictionary of Active Principles of Plants, 1894") has this formula . . . . . . $C_{34}H_{48}O_9$

Let it be further noted that Masson's formula for Bryogenin is .... $C_{14}H_{19}O_2$

The foregoing *Formulae* show that the exact chemical formation of *Bryonin* is not yet a matter of chemical certainty.

To proceed further, Dr. Dymock and his colleagues say that, in addition to *Bryonin*, they have not been able "to find a second bitter principle," in the tubers of *Corallocarpus epigaea.* I must here give Dr. Dymock and his colleagues, Dr. Wardell of Calcutta and Mr. David Hooper of Ootacamund, the fullest credit for their valuable researches. They go far ahead of even a more recent collaborateur of English repute, Mr. Charles E. Sohn, F.I.C., F.C.S.* Mr. Sohn merely rests content with an observation under the head of *Bryonin*, that it is a glucoside; amorphous; very bitter; yields sugar and *Bryonetin*" (*N.B. — No formula is given for this.* - K.R.K.); that "*Bryonin* is soluble in water, alcohol (2 to 3 parts) and insoluble in ether;" and that its precipitants are Tannic acid and Plantinum Chloride." In his *Addenda* Mr. Sohn notes that both *Bryonin*, and *Bryogenin*, as formulated by Masson, are both *dextro-rotatory*; the *Bryogenin* (which, be it noted, is a *glucoside*. — K.R.K.) dissolves with red colour in concentrated Sulphuric acid, "the solution becoming purple on heating, and giving, on subsequent addition of water, a purple precipitate."

That *Bryonin* is precipitated by Tannic acid was pointed out by Professor Dragendorff of Dorpat (Russia)† on the authority of Continental and American investigators. His formula for *Bryonin* is $C_{48}H_{80}O_{18}$ as noted above.

It may be useful to quote here, *in extenso*, a brief note I have come across in Vol. XXI of the "Pharmaceutical Journal," London, (1890 91, p. 496), headed "Constituents of the Root of *Bryonia dioica*," and embodying, as I may venture to say, the most recent Continental addition to the literature on the active principle of the entire genus known as *Bryonia*. The note runs as follows — "A treatise on this subject has been published at Dorpat by A. Mankowsky. He states

* Vide Sohn's "Dictionary of the Principles of Plants etc.," column 1, p. 23, 1894, London.

† Plant Analysis, "translated from the German by Henry G. Greenish, F.I.O., London, 1884, p. 170

therein that of the two glucosides contained in *Bryonia alba*, *Bryonin* and *Bryonidin*, the former is entirely without action, while the latter is poisonous only in large doses. The substances, hitherto known as *Bryonin* and regarded as the active agent of *Radix bryonioe*, are simply more or less purified extracts of the root, and probably contain both glucosides. The *Bryonin* of Schwerdtfeger is a mixture of the two glucosides, with the other substances, while Walz's *bryonin* is probably a mixture of this glucoside, with *bryodin* in smaller quantities. Introduced into the stomach, bryonidin causes inflammation of the stomach and larger intestines; introduced into the veins, only slight inflammation of the latter. The pancreatic juice decomposes *bryonidin*, and renders it inoperative." *Bryonidin* has no effect on the peristaltic action of the intestines, nor on the activity of the heart, nor on the peripheral nerves; introduced into living organs, it causes an enlargement of the vessels."

If any testimony were needed to establish the poisonous properties of the plant I am describing, the quotations I have cited above will be ample. Brigade-Surgeon Lyon was quite justified in drawing attention to *Corallocarpus epigaea* in his standard work on Medical Jurisprudence (vide p. 201, Edition 1889) while dwelling on the poisonous properties of some of the members of the *Cucurbitaceous Family*. Dr. Lyon says that the root of *Corallocarpus epigaea* contains a bitter yellow principle which Dr. Dymock suggests may possibly prove identical with *bryonin*. This Dr. Dymock and his colleagues have established in their subsequent researches and these I have quoted above. Sowerby and the Johnsons, in the second edition of their *British Poisonous Plants* (London, 1861, pp. 16-17), have very largely dwelt on the poisonous properties of the "Red-berried, common or white-bryony," — an English congener of the plant I am describing, and the only poisonous representative belonging to the *Cucurbitaceae* found among the British wild plants. Sowerby observes that "the large fleshy pale- coloured root is often seen suspended in herb-shops, occasionally trimmed into a human form." Remember my warning words in a foregoing page to the inexperienced Indian collectors of the mandrake root. The root, says Sowerby, is a powerful and highly irritant purgative, decoction of which is sometimes recommended by self-constituted medical advisers (of which, in India, likewise there is an endless number.— K.R.K.), "but dangerous to patients." Sowerby further adds that "the berries have produced vomiting when eaten by children." Our Indian children, half-starved though they are, and always in quest of field and forest fruit, never touch the fruit of the Indian species of Bryony I am describing. The fruit literally rots and dehisces on the plant. But there is one plant belonging to the *N.O. Menispermaceae* known as *Calumba*, and used in this country very largely and usefully too for medical purposes — the *Jateorhiza calumba* of Miers, formerly known as *Cocculus palmatus*, the commercial parcels of which, said O'Shaughnessy more than fifty years ago,* are adulterated *with Bryony dyed with turmeric.*" It is but natural that, when *Calumba* is so adulterated and sold in Indian bazaars, guileless and needy sufferers, having recourse to the adulterated root, under the guidance of the self-constituted medical advisers spoken of by Sowerby, should suffer from symptoms of violent vomiting and diarrhoea. No medical man prescribing *Calumba* ever anticipates such results. I have prescribed *Calumba* as a powerful tonic in dyspepsia, and as an alterative, especially in skin affections due to deficient action of the skin, but I have hitherto not had the experience of seeing any of my patients suffer from the effects of violent vomiting and diarrhoea when undergoing a course of *Calumba* treatment at my hands; for, knowing full well that the druggist and the herbalist often adulterate the *Calumba* root with turmeric dyed roots of the Indian species of Bryony, I have always taken the precaution of selecting the proper root. The root of *Calumba* is "perforated in the centre," says O'Shaughnessy, and "is met in slices of one or two inches diameter half-an-inch thick, much wrinkled, greyish externally, bright yellow within." It is not at all difficult for a careful observer to differentiate from the slices of the true *Calumba* root, slices of the Bryony root, dyed with turmeric. Even although in size. — *i.e.*, in diameter and thickness — the slices of the root of Bryony may exactly correspond with those of the *Calumba* root, note that the Bryony root *has not the central perforations spoken of by O'Shaughnessy* as existing in the *Calumba* root; then again the external coat of the *Bryony* root

* Vide p. 196, Bengal Dispensatory, Calcutta, 1841.

is yellow, having been tinged with the colour of turmeric, while it is greyish in the real *Calumba* root.

The reader will ask "why this digression?" "You are speaking of the poisonous properties of the *Bryony* root; what have you to do with the *Calumba* root ?" My answer is simple. I have already said, on high authority, that the root of *Calumba* is adulterated with turmeric-dyed bryony root; I have also referred to my own experience that the true *Calumba* root never produces vomiting or diarrhoea. But on referring to the article on the *Calumba root* in Dr. Dymock's "Pharmacographia Indica,"* I find very much to my astonishment that the experiments therein referred to as having been made by M. Houde (Repertoire, March, 1886, p. 113) and as resulting in M. Houde's being able to separate a *Crystalline principle, energetically productive of diarrhoea and vomiting* (*N.B.* — The italics are mine — K.R.K.), are puzzling unless it be supposed that the *Calumba* root was examined in an adulterated form. Because on the authority of Charles E. Sohn†, I am able to say that the two active principles hitherto found in the *Calumba* root — *i.e.*, *Calumbin* and *Calumbic acid* — are amorphous. It is only "from acetic acid," says Sohn, that *Calumbin* "crystallizes in prisms." He attributes no poisonous properties to it, such as noted by Houte. Houte's specimen, as analysed by him, must therefore be considered to be an adulterated one. It certainly could not have been adulterated with the root of any of the *Bryonias* as their active principles are glucosides, and none of them crystallized. It is possible some other tuberous root yielding a poisonous crystalline principle was adulterated with it, or sold in the market in the place of the true root of *Calumba.* The "*tricks of trade*" are unending.

The *Calumba root* is such an invaluable tonic that any adulteration of its root, with roots of a poisonous kind, likely to destroy its high medicinal value as a perfectly harmless tonic, and tending to credit it with not only *purgative* but even *emetic* properties, either of which properties as a matter of fact do not exist in it, should be jealously guarded. This I have ventured to do as a side-issue of this paper; for I know, and O'Shaughnessy supports in my experience, that the

* Vol. I, p. 47, Bombay, 1890

† P. 24, Edition 1894, London, "Dictionary of Active Principles of Plants," & c. Sec. also page 116 where Columbin is said to be "Amorphous" in column 4.

Calumba root, unadulterated, seldom if ever occasions any unpleasant sensation in the stomach or alimentary canal, but *When adulterated* with the turmeric dyed root of Bryony, it is, not unoften, administered, though unknowingly, with fatal results. Such has been my painful experience.

Remember here that *Bryoidin* referred to by Professor Dragendorff under head "*Fixed oils, &c.*" (Plant Analysis - English Translation by Greenish, p. 109), has no connection whatsoever with the active principle of the Bryony root. For according to the researches of Professor Fluckiger, *Bryoidin*, constitutes one of the remarkable series of the *Elemi* constituents. "It is a crystallizable substance, *occurring in that resin*, only in small quantities." Be it noted further that this organic substance known as *Elemi* is "a concrete resinous exudation, the botanical source of which is still undetermined but sometimes referred to *Canarium commune* (or to *Icica alba*), Manilla."*

It may be useful at this stage to quote the other authoritative English writers who have expressed themselves on the poisonous nature of the Bryony root.

Loudon has observed that the root of the genus *Bryonia* "is a famous hydragogue, and highly purgative and acrid."†

Beck has noted that the administration of *Bryony root* "has caused vomiting, purging, fainting, violent, pain, and profuse alvine evacuations."‡

James Aphjon includes the English representatives of the genus *Bryonia (alba* and *dioica*) among plants "which yield the more important vegetable acrids.§" The acrid virtues of Bryony, says he, "reside in a principle soluble in water which was discovered not long since by Brandes and Firnhaber, and which they have named *Bryonine*. There are upon record some cases in which decoctions of the *Bryony* produced death with hypercatharsis." It may be stated here that the root of the Indian congener I am describing is capable of doing it.

* Lander Brunton's Pharmacology, &c., Edition 1885, London, p. 816

† Encyclopædia of Plants, 1829, p. 811

‡ Medical Jursprudence, p. 829, 5th Ed., London, 1836

§ Vide Article " Toxicology," Vol. IV, p. 219, Cyclopædia of Practical Medicine by Drs. Forbes, Tweedie and Conolly, London, 1835.

Dr. Alfred S. Taylor, the sheet-anchor of modern Forensic talent, says that the powdered root of *Bryony, black or white* (note that with the *black- bryony* we are not concerned at present — K.R.K.) when swallowed produces severe pain, vomiting and purging, and after death the stomach and intestines are found highly inflammed.*

It is somewhat disappointing to find that in his excellent "Elements of Pharmacology," Professor Oswal Schmiedeberg, of the University of Strasburg, has made no allusion to the active principle of *Bryony*.

Baillon† makes the following remarkes regarding *Bryony* — "In our common *Bryony* (*Bryonia dioica*) the subterranean portion — white, poisonous, caustic - is extremely irritating to the skin and mucous membrane. It is a powerful dangerous purgative, prescribed chiefly in dropsical and various nervous affections. Freed from its irritant principles, this portion is rich in starch which has been turned to account, and from which alcohol is distilled."

The testimony of Colonel Drury‡ with regard to the Indian representative of the genus *Bryony* —namely, *Corallocarpus epigaea* or *Bryonia epigaea* as he describes it, is absolutely worthless. His remarks show that his information is only second-hand and as such not of much value.

The English *Bryonia dioica*, says Lauder Brunton,§ "is made into a tincture (bitter principle *bryonin*) the chief use of which is that of a *hydragogue cathartic*," but he adds "it is now superseded by Jalap," which latter I may say is certainly safer and less irritating to the *primoe vioe*.

Here is Sir Robert Christison's testimony regarding the poisonous nature of the entire genus *Bryony* :|| "Before *Bryony* root was expelled from medical practice, it was often known to produce violent vomiting, tormina, profuse watery evacuations and fainting. Pyimentions a fatal case of poisoning with it which happened at Cambray in France. The subject was a man who took two glasses of an infusion of the root to cure ague, and was soon seized with violent tormins

* Alfred S. Taylor on Poisons, page 510, 1848

† History of Plants, p. 421, Vol. VIII, London 1888

‡ The useful Plants of India, 2nd Edition, 1873, p. 88

§ Pharmacology, &c., Ed. 1885, London, p. 848

|| P. 594, Edinburgh, 1845, A Treatise on Poisons

and purging, which nothing could arrest, and which terminated fatally (*Neues Magazin*, I. 3, p. 557). *Orfila* quotes a similar case from the Gazette de Sante which proved fatal within four hours in consequence of a strong decoction of an ounce of the root having been administered partly by the mouth and partly in a clyster to repel the secretion of milk. (*Toxicol Gen.*, I, 680)."

Let this unquestionable testimony of Orfila, quoted as it is by such an equally unquestionable authority as Sir Robert Christison, be a warning to those who delight to call themselves the admirers and followers of *Homoeopathy*. I have recently come across a writer on the "*Homoeopathic use of Plants*"- Richard Hughes, L.R.C.P., Ed. — who, in an elaborate volume entitled A Manual of *Pharmacodynamics* (5th Ed., Lond., 1886, p. 334), makes the following remarks regarding the action of *Bryonia* on the *mammary glands* — "Whenever from the first coming in of the milk, from catching cold, while nursing, or from abrupt weaning, the breast becomes swollen, tender, knotty, and painful, *Bryonia* will almost certainly resolve the inflammation and prevent the formation of abscess. Dr. Durham advises it in milk- fever." As against this homoeopathic preaching, compare the testimony of our leading Allopathic Physician, Sir Robert Christison's reference to Orfila's experience, as cited above, that the administration of the *Bryony* root, partly by mouth and partly by clyster, "*to repel the secretion of milk*," has been followed by fatal results. So much for the glory of *homoeopathy* and its admirers !

Having so far dwelt on the European writers describing the poisonous properties of the genus *Bryony*, I may briefly state what Indian writers have said regarding them.

Dr. Edulji Cowasji Appu (Tukina), L.M. and S., of the Bombay University, says emphatically, in describing the properties of the genus *Bryony*, that its root contains "a peculiar bitter principle named *Bryonin*, which in non-medicinal doses is highly poisonous, acting as a violent emetic and cathartic, giving rise to symptoms resembling those of cholera."* I can safely vouch for this statement.

Shodhal, an old Sanskrit writer, has distinctly noted that plant is *hydrogogue* and *emetic*.†

* Practitioner's Vadc Mecum, p. 120, 1891, Bombay

† Vide Katâ Bhat's Nighant-Sangrah, Junagadh, 1893, p. 569

Jaikisan Indraji says that the odour of the root is acrid, but this is a mistake. The root has no odour. It is the taste that is bitter and acrid; mucilaginous and slightly acid.

Dr. Sakharam Arjun* and Dr. Virji Zina Raval† dwell on its medical properties only. They both make no reference to its poisonous properties.

To sum up, whether the poisonous element is *Bryonin* as Dr. Tukima says, or whether it is *Bryonidin* which causes inflammation of the stomach and larger intestines as mentioned above in a quotation from the Pharmaceutical Journal (1890-91, p. 496), there is scarcely any doubt but that the root of the *Bryony Genus* may be safely looked upon as a powerful irritant of the entire elementary canal, causing dangerous vomiting, accompanied with fainting, and almost fatal hypercatharsis.

O'Shaughnessy states that the powder of the root of the plant I am describing, which he, according to the nomenclature of his day, calls *Bryonia epigaea* is given by native practitioners as an aperient and alterative in chronic dysentery and venereal complaints. —(Ainslie, vol. II, p. 158). I am afraid both Ainslie and O'Shaughnessy are trusting to second hand information, else Ainslie at any rate would not have committed himself to the somewhat dangerous statement that such a dangerous irritant of the *primae viae* could, by any possibility, be a remedy in chronic dysentery. In venereal complaints it may be a possible remedy, *but on other physiological grounds*, which I need not discuss here, it is by no means a safe remedy.

In concluding this notice, I may here note that the Sanskrit synonym *Kadamba* given to *Bryonia epigaea* in the Mangalore book, issued under the title "*Five Hundred Indian Plants*," ‡ is positively wrong. I can say with confidence that the Sanskrit plant known as *Kadamba* does not belong to the *Cucurbitaceae*, but to the *N.O. Convolvulaceae*. Burnouf gives the botanical name of the Sanskrit term *Kadamba* as *Convolvulus repens*. (*Vide* Sanskrit and French Dictionary, page 136). The *Kadamba* of the Dekkan and Konkan belongs to *N.O. Rubiacioe*.

---

* Bombay Drugs (Catalogue)

† P. 168, Ahmedabad Edition, 1889. Arya-Aushadh.

‡ Five Hundred Indian Plants : Their uses in Medicine and the Arts, in Kanarese; p. 11, 1891.

Reprinted from
Journal Bombay Natural History Society
Vol. 10(4) : 618-627. 1895

PART XVI.

PLUMBAGO ZEYLANICA — (*Linn.*)

Reprinted from
Journal Bombay Natural History Society
Vol. 10(4) : 618-627. 1895

PART XVI.

PLUMBAGO ZEYLANICA — (*Linn.*)
Natural Order — PLUMBAGINEAE.
MARATHI. — चित्रक (*Chitrak*).

An evergreen perennial herb or undershrub, growing in gardens, but not wild, in Thana.

ROOT. — Long, succulent, often much contorted and substantial.

STEM. — 2-5 feet, spreading, cylindrical, glabrous, striate, often much branched.

BRANCHES. —Diffuse, striated longitudinally. Length of the branches, often 4-5 feet. Procumbent branches, on reaching the soil, often strike root.

LEAVES. —Alternate, exstipulate, entire, sessile; $1\frac{1}{2}$ - $3\frac{1}{2}$ inches long, 1 inch broad, oval or ovate-oval; "much tapering towards the base, but there dilated to form rounded, amplexicaul, stipule-like auricles, acute at apex, undulate, glabrous, thin, somewhat glaucous, and finely scurfy beneath" — (Trimen).* Clarke says the leaves are puberulous† or glabrous, acutre, 3 x $1\frac{1}{2}$ inch.

PETIOLE. — 3/4 inch; base dilated, amplexicaul (Clarke).

VERNATION. — Involute.

INFLORESCENCE. — Spikes 4-12 inches long; often branched.

FLOWERS. — Seentless, white, regular, bisexual; "on very shot peduncles, in erect spicate terminal racemes; rachis and bracts glandular" (Trimen). Bracts and Bracteoles, persistent, leafy, shorter than the calyx.

BRACTS. - Though shorter than the calyx when the flowers mature and expand, they usually sheath the flower buds completely in the

* Trimen's Handbook of the Flora of Ceylon, Vol. III, p. 65, 1895.

† C.B. Clarke's article in Hooker's "Flora of British India," Vol. III, p. 481.

early stage of floral development. They are about 1/5 inch long, ovate and acute. They are "studded with glands" as Trimen observes. The nature of these glands will be dwelt on further.

BRACTEOLES. — Subulate; 1 or 2; smaller than the bracts and distinctly glandular.

CALYX. — Inferior, free 1/3-1/2 inch by 1/10 inch, persistent, narrowly tubular, or spindle-shaped; "5-10 ribbed, often hyaline between the ribs; mouth frequently funnel-shaped, scarious" (Clarke). Segments, five, with membranous margins; "covered with large, stalked, spreading, globose, crimson glands" (Trimen). "Teeth very short" (Clarke).

COROLLA. — Inferior, persistent, tubular. Tube "fully 1 inch long and slender," says Trimen; "3/4 inch," says Clarke. I may add that this is a matter of purely local development, and does not count for much. Petal-lobes, 5, spreading, nearly 1/2 inch long, or more frequently less, oval, ovate, or ovoid; acute.

ÆSTIVATION. — *Twisted* in the case of petals; *plicate* in the case of the calyx.

It may be noted here that the corolla is less persistent than the calyx. This is characteristic of the genus *Plumbago*.

STAMENS. — 5, hypogynous, distinct, opposite the petals.

FILAMENTS. — Free; as long as the corolla-tube; linear, dilated at the base.

ANTHERS. — Oblong; exserted just beyond "*the throat*".

HYPOGYNOUS.— DISC. - Absent (Clarke).

PISTIL. —

OVARY. — Superior, 1-celled, 5-angular, narrowed at the apex.

STYLE. — Slender, simple, filiform; divided into five stigmatose branches nearly throughout the length. "Base of style, glabrous" (Clarke).

STIGMAS. — 5, capillary and quite distinct.

OVULE. — Solitary; pendulous from a basal ascending funicle.

FRUIT : A membranous capsule; oblong, sharply-pointed "included" in a persistent calyx and corolla. "Pericarp, thin below, thick and hard above" (Trimen).

DEHISCENCE. — Circumsciss near the base.

SEED. - Cylindrical or oblong-rotund, with undefined longitudinal striae; green when immature; brownish when mature

ALBUMEN. — Farinaceous.

EMBRYO. — "In axis of the endosperm" (Trimen).

RADICLE. — Superior.

COTYLEDONS. - Flat.

PLUMULE. — Slightly rough.

This description is mainly based on Trimen's and Clarke's in works cited in the foregoing foot-notes, coupled with my own observations of the plant as grown in my Thana garden from year to year for over fifteen years.

REMARKS.

Much of what I should have liked to note under this head has already appeared in the previous pages of this Journal (*Vide* pp. 349-357, Vol. IX, No. 4) in my description of *Plumbago rosea* — a congener of the plant I am now describing. My work, therefore, under this head is much curtailed.

The plant has the following synonyms —

1. *Plumbago auriculata.* — Blume, Bijd, 736.

2. *Pl. flaccida* — Moench, Meth. 429. (*Kew Index* — 2nd Column, p. 569, Vol. III, 1894).

3. *Pl. viscosa* — Blanco, Fl. Filip. Ed. I, p. 78. (*Kew index* — Volume and page cited as above).

4. *Pl. sarmentosa* — Lam. Illustr. I. 470, *Scandens*. (*Kew Index.* — Volume and page cited as above).

5. *Thela alba*, Lour. Fl. Cochinch 119. — Rheede Hort. Mal., X, t. 8 (*Hooker's "Flora, Br. Ind.," Vol.* III, p. 480).

As noted above, Lamarck gives the plant the name *Plumbago sarmentosa*, which is a *scandent* variety of the plant. Under suitable circumstances, such as a rich soil or proper manuring of leaf-mould, I have known the plant to deviate from its suffruticose habit, and assume the scandent form.

The leaves, according to Roxburgh*, are "ovate, suddenly narrowed into the petiole; rachis of the spike pubescent, or glandular."

The bracts are "threefold," says Roxburgh, and "one-flowered; the outer bract is ten times larger than the lateral one; they are covered with the same gluten as the peduncle of the raceme; sometimes there is a fourth linear bract pressing the calyx."

With regard to the special characters of the glandular bodies found on the various parts of the plant under description, I must referthe

* "Flora Indica," Vol. 1, p. 462

reader, as I did in a paper published in a former volume of this Journal,* to the elaborate researches of Dr. John Wilson of the University of St. Andrews, contributed to the pages of the "Annals of Botany,†" in an able paper entitled "*The Mucilage — and other glands of the Plumbagineae.*"

Dr. Wilson describes two sets of glands, *viz.* — (1) The *Mettenian glands*, which in certain species of the *Plumbagineae*, or under certain conditions, secrete chalk- scales, (2) *Mucilage-glands*, which secrete a viscid mucilage. These glands are of two kinds. — (a) *Stalked* external glands; (b) minute internal glands which are *sessile*. These latter are found on the calyx of *Plumbago*.

(1) With regard to the *Mettenian glands*, Dr. Wilson observes thus — "In general, the glands occur in the longitudinal depressions on the stems of plumbagos, giving rise to a striated appearance, where calcium carbonate is secreted." They are present "on the leaves and bracts of all, and on the sepals of many of the species" of *Plumbago* and of other genera of the natural order *Plumbagineoe*. They are known to occur on the cotyledons of *Plumbago*, "even before the light is reached by the germinating seedlings," says Dr. Wilson. In his observations on the seedlings of *Plumbago zeylanica*, this careful observer notes thus — "Examples of different ages were closely examined, revealing the fact that no mucilage-glands exist at the base of the cotyledons. Mettenian glands are as usual present on the cotyledons, and they likewise occur sparingly on the hypocotyledonary axis. A seedling which had the plumule destroyed at an early stage continued to vegetate by its cotyledons alone, which persisted for many months. They were ultimately 1 5/8 in. long and 5/8 in. broad, and presented the very noteworthy phenomenon of bearing, like the leaves, chalk- scales on their undersurface."

In my remarks on *Plumbago rosea* in a former number of this Jr. (*loc. cit.*), I candidly confessed that for a long time I did not know whence came the chalk-scales or the white crust on the leaves of *Plumbago rosca*; I now say that the same was my difficulty in the case of the white crust on the organs of *Plumbago zeylanica*. It would be quite appropriate to observe here that even such a consummate botanist as Professor Vines Cambridge has fallen into error in tracing

* Vol. IX, pp. 345-365. *Plumbago rosea*, Linn.

† Vol. IV, pp. 231-255; 1889-1891.

the source of this calcareous deposit on the leaves of the Plumbagos. He traces the origin of the white crust to the epidermal cells. But Dr. Wilson points out the error of Professor Vines in words which are well worth reproducing here. They are as follows — "It may be remarked here that Vines* is misleading in the following statements — 'Similar scales (Calcium carbonate) occur also on the leaves and herbaceous stems of various plumbaginaceous plants. In these cases no glands like that described (water gland) are present; it is, therefore, to be concluded that *the epidermal cells themselves* excrete the calcium carbonate."

(2) With regard to the *Mucilage-glands* noticed by Dr. John Wilson in the Plumbagos, I have this to observe — Dr. Wilson says that in *P. zeylanica* the mucilage-glands are of the same size as those in *P. rosea* and P. Capensis, the largest being 38 mm. in diameter. In the specimens of *P. zeylanica* examined by Dr. Wilson, he says — "A large number were seen clustered in the axil of a young leaf." Dr. Wilson gives a graphic figure in his paper, illustrating the relative sizes of the basal and secretory cells of one of these glands, which he modestly calls "a somewhat small example," but which I think is sufficient, for all practical purposes, in understanding the primitive structure of the mucilage-glands of the entire order *Plumbagineae* Hitherto unknown in the literature of British Botany. To come to the description of the first sub- division of the *Mucilage-glands, viz.* :- (a) *the stalked external glands*; they are seen outside the calyx and on the rachis in *Plumbago zeylanica.* Dr. Wilson says that in *P. zeylanica* the entire rachis is thickly studded with active glands, like the calycine ones, but more minute. The second sub-division of the *Mucilage-glands*, includes (b) *the internal sessile glands.* They occur on the inner face of the sepals and are to be seen in *P. zeylanica.* "Their basal cells are usually four in number," says Dr. Wilson, "but very often two (*P. rosea*)." He adds that the basal cells "in all cases support numerous very thin-walled secreting cells." The sessile glands, be it noted on the authority of Dr. Wilson, "do not offer any point of material distinction from normal mucilage glands of the leaves." Such is a brief resume of the observations of Dr. Wilson regarding the glandular appendages of the plant I am describing. I have tried to reproduce here every observation of Dr. Wilson on this very interesting question

* Lectures on the Physiology of Plants, P. 246, Cambridge, 1886. (J.W.)

of the glandular appendages of *P. zeylanica* for the benefit of my readers. Superficial readers will find it irksome, but to the practical student of Botany, Dr. John Wilson's researches are of rare value, this consideration is my only apology for thus occupying the pages of this Journal.

With regard to the presence of pith in *Plumbago zeylanica*, I may notice this fact that the new branches which shoot from the root direct, as apart from the main-stem, on cross-section show a good deal of pith in the central part. This morphological condition does not exist in the regular branches of the main- -stem. The longitudinal striation of the green-glaucous bark is also well-marked in the branches shooting forth from the base of the stem.

The flowering time given by Trimen in Ceylon is between December and March. In September last the plant flowered in my garden in Thana, and now in the middle of October there is no flower; the fruit is maturing. The last time I saw my garden plant flower was in February of the current year. It may therefore be safely conclude that the plant flowers twice a year on this side of India in gardens.

As a "Stove-evergreen" in England, the plant flowers in June (Johnson)*; Paxton† says the same. He gives the height of the plant as 1 1/2 ft.; but I have in my garden, under the Indian sky, branches from the main-stem measuring quite a yard in length, and exhibiting a scandent habit. The plant appears to have been introduced into England so far back as 1731. This is a point of historical interest.

Kurz is brief in his description of the *Plumbagineoe* in his forest. Flora of Burma (1877, two volumes), regarding the plant I am describing. He disposes of the plant by simply saying it is a "vesicatory (p. 96, Vol. II).

Dalzell and Gibson‡ observe that the divisions of the "limb", the corolla are "cuneate and retuse." They remark also that the plant is "common on the rocky places in Concans." I do not know if it is meant hereby that the plant grows wild in the Concans. I should like to know whether this plant is wild on the side of India, or whether it is an introduced plant, as the name indicates, from Ceylon, subsequently cultivated in gardens and the

* Johnson's Gardener's Dictionary, Ed. 1894, p. 772

† Paxton's Botanical Dictionary, Ed. 1868, p. 449

‡ Bombay Flora, p. 220, Ed. 1861.

almost naturalized. I say "almost naturalized," as it appears to take kindly to any garden-soil in and around Bombay. Dalzell and Gibson conclude their notice of this plant with the somewhat puzzling remark that the plant "grows from Cabool to New Holland." Is it meant here to say that the plant grows *wild* from the capital of Afghanistan right down to the Austral regions we now know as Australia? Or is it meant to say that the plant is cultivated in garden throughout the tract between Cabool and Australia? Would some better-informed Oriental Botanist come forward to elucidate this point ?

Although as I have said above, Kurz dispose of the description of *P. zeylanica* in a few words in his "Forest Flora of Burma" (1877), it evidently not being a forest plant of any pretension whatsoever; in a communication to the "Journal of the Asiatic Society of Bengal," which he published the same year, either simultananeous with, or subsequent to, the publication of his "Forest Flora of Burma," he makes the following remark as regards the *habitat* of the plant — "In rubbishy places in and around villages, along river-banks and *toungyas,** not unfrequent all over Burma, but apparently nowhere really wild.†" Such is my own experience on this side of India, namely, that the plant is not wild here, but a cultivated one, which has taken kindly to the garden areas of our province.

Perhaps it would be useful to transcribe here the main features of *Plumbago zeylanica* from the original Latin, expressed in his *Hortus Malabaricus* (Vol. X, page, 15, t. 8). "Height of the plant, 2-3 ft., growing in sandy soil; root, yellowish black, with hairy roots." This hairy condition of the root I have been able to recognize in many garden specimens of fresh root; but I believe this is due mainly to poor soil- nourishment, for, as observed before, in most of the specimens I have examined from Thana and Bombay gardens, *Plumbago zeylanica* (Linn.) furnishes a root which is distinctly "long, succulent, often much contorted, and substantial" (K.R.K.). I hold myself soley responsible for this statement, and I would request every Bombay Botanist, who has specially studied the plant as I have

* *Toungya = Toung-ya*, which in Burmese means a form of hill-cultivation carried on by bouring the jungles (Balfour). See my note, p. 352, Vol. IX, Bombay Natural History Journal, 1895.

† S. Kurz - Contributions towards the knowledge of the Burmese Flora, page 217, Part II (Natural History &c.), vol. XLVI, Calcutta, 1877 (Nos. I-VI).

had to do, to be able to write this paper after close study, either to contradict or verify my assertion. I am quite prepared to accept any view. Rheeded says that the colour of the flowers is "completely white" (Latine — *Coloris in totum candidi*). Rheede wrote his description over two hundred years ago, under a different sky — more southerly in direction. I mention this particularly by way of an apology as my plate, painted by the skilful hand of Mr. Isaac Benjamin from nature under my direct supervision, has a faint dash of the azure tint, not far removed from the light blue of *Plumbago capensis*. The natural size of the illustration (Plate R. accompanying this letter press) will be amply sufficient to show that even if *Plumbago zeylanica*, at times under climatic influences, assumes a slightly bluish tinge, it can never approach the pretty bluish tinge — more strongly marked in *Plumbago capensis*. The latter is altogether a larger-flowered plant, most easily differentiated from *Plumbago zeylanica*.

The root of the plant requires some species remarks, as that is the part which is mostly used for criminal purposes. A full description of the dry and fresh roots is given in the "Pharmacographia Indica,"* which it would be quite appropriate to quote here. It runs as follows : "The roots of *Plumbago zeylanica* are from 1/4 to 2 or more inches in diameter, seldom branched. When dry, the external surface is of a dark reddish-brown colour, somewhat shrivelled, and marked here and there by small warty projections; internally it is brown and striated; the fracture is short, the taste acrid and biting." Let me observe in passing that Rheede says (*opus cit.*) that it is less acrid than the root of *Plumbago rosea* (Tambido-Citroko, i.e., the red Chitrak), — (K.R.K.). To continue the remarks of Dymock and his collaborateurs (*Opus cit.*) as regards *Plumbago zeylanica*, they oberserve thus : "Wood hard, reddish, close grained. A section of the fresh bark, when magnified shows numerous bundles of bright yellow stone-cells forming an irregular zone towards the inner part of the middle layer of the bark. The cells of the parenchyma are large and contain much starch. In the dried root the yellow *Plumbagin* is seen in the cell-walls of both the parenchyma and the woody tissue, but not in a crystalline form (*Fluckiger* and *Gerock*)."

* Vol. II, p. 331 (Dymock, Warden and Hooper). "Phamacographia Indica", 1891, Bombay Education Society's Press.

POISONOUS PROPERTIES.

What I have said of the poisonous properties of *Plumbago rosea*, applies equally to the plant I am describing. Dymock and his colleagues, however, say in their "Pharmacographia Indica,"* that "*Plumbago zeylanica* is mentioned by several European writers upon Indian drugs, but has not attracted the same amount of attention as *Plumbago rosea, which is said to be more active.*" (The italics are mine. — K.R.K.) Dr. Dymock says that Plumbago root given internally in large doses acts as a *narcotico-irritant* poison.

O'Shaughnessya† attributes to the root vesicant properties, not unlike cantharides when used *externally, i.e.*, over the skin. To quote his very words, the properties of *Plumbago rosea*, "and of the *Plumbago scandens* and *Plumbago zeylanica* are nearly identical. It is applied by the natives of India to buboes in the incipient stage. In fact, says he, the prevailing characters of the entire Indian *Plumbagineoe* are "astringency and acridity."

Udoy Chand Dutt‡ says distinctly that the root of the *Plumbago rosea* is a more powerful vesicant than that of *Plumbago zeylanica*. That may be; but I do not suppose there is any difference between the two as regards their abortifacient property. They contain the same active principle, *Plumbagin*, which I have already dwelt on in my paper on *Plumbago rosea*. Dr. Dymock observes§ that "at present the physical properties of the principle, the so-called *Plumbagin*, are not sufficiently well-known to enable one to positively assert whether it is odourless or not, while its chemical constitution has not been studied." It is, however, a matter of certainty that the active principle is detactable in the contents of the stomach by the action of subacetate of lead. We have the authority of O'Shaughnessy on this point. This is what he says — "The root, in various forms, is much employed as a poison in India; and, as an irritant to occasion abortion, it is introduced into the vagina and applied directly to the neck of the uterus. In a criminal case of the former kind in 1837, the Editor ¶ succeeded in detecting the poison by acting on the contents of the stomach with alcohol, concen-

* Vol. II, p. 330, (op. cit.) 1891, Bombay

† "Bengal Dispensaory", pp. 508 and 509, Ed. 1841, Calcutta

‡ The Materia Medica of the Hindoos" Calcutta, 1877, p. 187.

§ Pharmacographia India, p. 334, Vol. II, 1891

¶ i.e., Dr. O'Shaughnessy himself - K.R.K.

trating the tincture, redissolving in a small quantity of boiling water, and adding the subacetate of lead, by which the very characteristic red colour was immediately occasioned. By a modification of this process two grains of the powdered bark may be detected in a pint of a mixture of milk, blood and various articles of food."* Dr. Lyon classes *Plumbago zeylanica* among *irritants* and observes that "powerful irritants may like purgatives cause abortion, owing to the uterus participating in the irritant action set up in the system†." Mr. Gribble, late of Madras, and now of Hyderabad, Dekkan, in describing the poisons most generally used for criminal purposes in the Madras Presidency, refers to *Plumbago zeylanica* thus — "The use of this poison is by no means uncommon; five cases occurred in 1882 and three in 1883‡." Mr. Gribble classes the plant among *irritant* vegetable poisons which may be identified by chemical or by physiological tests (p. 225, *op. cit*). Dragendroff notes that *Plumbagin* is coloured cherry-red by small quantities of alkalies. He includes it among the bitter principles soluble in ether ("Plant Analysis," p. 146-149, ed. 1884). The table given in the "Pharmacographia Indica" (pp. 335-339, vol. II), showing the particulars of plumbago-poisoning in India, is well worth the perusal of those wishing to have more information on the subject. The following observation of Dr. Norman Chevers§ may be usefully quoted here : "Dr. Shortt|| remarks upon the use of *Plumbago zeylanica* or *Chittra moolum*, that "it is a universal remedy in dysmenorrhoea, and equally so as a means of producing abortion; there is perhaps scarcely a village in India where the use of this drug as such is not known. Its action is that of a powerful irritant; it is not only had recourse to as an internal remedy, but is more frequently applied externally." I gather from Dr. Norman Chevers, quoted cases (pp. 739-741, *op. cit.*), that, though plumbago may cause abortion, it does not necessarily cause the death of the mother; "some women escape with life after having aborted."

* Bengal Dispensatory p. 510, Calcutta edition, 1841

† Medical Jursprudence for India, p. 379, 1st ed. 1889. (See also p. 106, Vol. VIII, of this society's Journal for Professor Schmidberg's remarks in the connection, quoted by me - K.R.K.)

‡ Outlines of Medical Jursiprudence for Indian Criminal Courts, Madras, 1885, p. 201.

§ A Manual of Medical Jurisprudence for Indian Criminal Courts, Madras, 1885, p. 201.

|| Fœticide, (Madras Quart. Jour. of Medical Science, July 1861, p. 262), quoted from N. Chevers.

Reprinted from
Journal Bombay Natural History Society
Vol. 11(1) : 252-261. 1896

PART XVII.

NERIUM ODORUM — (*Solander.*)
Natural Order — APOCYNACEAE.
MARATHI — कण्हेर (*Kanher*).

A large erect stout glabrous shrub, containing a cream-coloured, thick, resinous, and sticky juice. The generic name of the plant is derived from Gr. *neros*, moist, which fully indicates its habitat. It thrives in damp ground, and alongside of flowing rivulets and watercourses. It is a beautiful evergreen, easy of culture and propagation, and flowers freely throughout the year.

ROOT. -Crooked on young roots the corky layer is very thin, & the interior yellow colour of the bark is seen through it; inner surface yellow. Odour somewhat acrid; taste acrid and bitter. In the bark of the roots, the medullary rays are very numerous, their being loaded with yellow resinous juice makes them very conspicuous (Dymock).

STEM. - 6-8 feet high, woody; abounding in starch; pithy in the centre. Young branches show more pith. Wood very porous and abounding in large dotted vessels and well-marked; medullary rays visible to the naked eye. The laticiferous vessels are numerous, and generally in groups of two, three, or more. (Dymock).

BARK. - Thick, soft, corky, with a grey surface externally. In young branches it is greenish; in old yellowish and abounding in starch.

LEAVES - Exstipulate; arranged in whorls of three; rarely opposite or scattered; narrow, linear-lane oceolate, 4-6 inches long, thickly coriaceous, acuminate. Margin entire, revolute. Midrib very stout; main lateral nerves numerous, slender, horizontal, parallel, very close. The under surface, says Brandis, is uneven, and irregularly pitted. The margin narrows as it reaches the petiole.

PETIOLE — Very short.

FLOWERS — Hermaphrodite, showy, sweet-scented; single or double, variously coloured, as detailed below; 1-1/2 inches in diameter : salver shaped.

INFLORESCENCE. — Racemose cymes.

PEDUNCLES. — Terminal, long, angular.

PEDICELS. — Short.

BRACTS. — Linear, deciduous, coloured.

ÆSTIVATION — Valvate in the upper part of the calyx tube; contorted in the corolla, segments overlapping to the right.

CALYX — Inferior, 5-partite, tubular persistant, slightly accresscent; segments subulate-lanceolate, erect, says Roxburgh. The base of the calyx-tube glandular within. Professor Vines of Cambridge, *apropos* of the Nectaries, observes thus :- "Nectaries many, in a whorl as rounded prominences, situated extra-staminally, between the corolla and calyx" ("Text-Book of Botany," pp. 526-527-1895).

COROLLA. — 5-lobed, twisted, hypogynous, gamopetalous, regular, deciduous. Appendages of the corolla cleft into 4-7 segments (Hooker); Brandis, however, says "that the corona of each petal is *trifid*; the lateral segment is linear, and the centre one short, triangular." This is the common arrangement in single flower, so far as I have observed. In the flesh-coloured flowers, however, the arrangement is somewhat different, in that the corona is more divided, having five strap-like laciniae, quite entire half-way up, and then uniformly dividing into a filiform fringe.

ANDRŒCIUM —

STAMENS. — 5, alternate with the lobes of the corolla, and included within the corolla-tube.

FILAMENTS — Distinctly attached to the tube the whole way down.

ANTHERS. — Sagittate, introrse, united to the stigma, 2-celled, dehiscing longitudinally. "Spurs of the anthers, linear, twice as long as the cells" (Solander). The *Connective* prolonged into a feathery process more than twice the length of the anthers. These feathery processes, five together, are spirally twisted into a bundle, which projects well beyond the tube of the corolla; the individual processes however are easily separable.

POLLEN. — Globose; immediately applied to the stigma.

GYNOŒIUM. —

OVARY — Superior, composed of two carpels, which are coherent, but easily separable in the fruit.

STYLE — Single, uniting the ovaries.

STIGMA — Hour-glass or dumbbell-shaped. Very characteristic.

OVULES. — Numerous, anatropous.

FRUIT. — A cylindrical capsule, with deep linear striations, slightly twisted, sometimes very much so; 6-9 inches long.

SEEDS — Numerous compressed, exalbuminous, comose, with a tuft of fine shining white or greyish silky hairs at the top; fusiform and slightly rugose (Gaertner).

EMBRYO — Straight - (De Caisne and Le Maout).

COTYLOEDONS. — Foliaceous, cordate, convolute.

RADIX. — Superior : slightly rounded.

Both the cotyledons and radix partake of the red or pink colour in the respectively coloured varieties (Gaertner).

GENERAL REMARKS

It must now be considered that the *Nerium odorum* I am here describing is no other than the *Nerium oleander* of the Mediterranean coast, barring developmental differences due to climatic influences. Linnaeus is after all right in considering that they were identical plants. However much the corolla may vary in the two plants, we have the high and unquestionable authority of Brandis, that the fact of a mere climatic variation of the corolla " does not afford distinctive characters of a reliable kind. Special parts may vary but yet thier variation need not go to multiply varieties which may reasonably be classed under one and the same species." This is so with *Nerium odorum*, notwithstanding its single and double- flowered white, pink, and crimson varieties, as will be seen from the synonyms I detail below on the authority of several well - known Botanists.

Synonyms —

1. *N. odoratum*, Lamk.
2. *N. latifolium*, Mill.
3. *N. indicum*, Mill.
4. *N. flavescens*, Spin., Jard. S. Sebast (1812). Kew Index.
5. *N. Koltschii*, Boss. Diagnosis Pl. Or. Nov., VII, p. 21, 1846, Kew Index.
6. *N. luteum*, Nios. Kew Index.
7. *N. verecundum*, Salisb. Kew Index.

The following garden varieties have been noted as growing in Bombay —

1. The double white (=*N. album plenum*).

2. Single white, a slightly cream-coloured (=*N. album*, or *N. flavescens*).

3. The flesh-coloured, or pale pink (=*N. odorum*, var. *carneum*. See plate 2032. Curtis, Bot. Magazine).

4. Double rich pink (=*N. plenum*). See plate 1799. Bot. Magazine. Syn. *N. indicum*, Mill; *N. zeylanicum*, Burm.; *N. latifolium indicum* Commel.

5. Single crimson — The species or variety described as *N. odorum*.

To students of Botany devoted to the microscopical examination of the morphology of the plant I am here describing, Professor Bastin of Philadelphia recommends the following parts as good examples for microscopic section* — (1) The corky tissue; (2) simple laticiferous tissue, in contrast with the complex, *i.e.*, articulated laticiferous tissue found in such plants as Poppy and Dandelion; (3) both the surfaces of the leaves show stomata.

Apropos of the above quotation, under the second head, may be considered a very cogent remark of Kerner's in his Natural History of Plants (translated into English recently by Oliver, Page 470, Vol. I), which is to the following effect — "There are many laticiferous tubes in the oleander." I may add that what is true in this connection of *N. oleander* is equally true of *N. odorum*, in all its varieties bearing variously-coloured flowers.

With regard to the general manner of branching in *N. odorum*, Kerner has very accurate remarks to make, borne out by my repeated observations of the severally coloured single and double-flowered varieties visible in and around Bombay. This is what Kerner says — "In the oleander, when after the fall of the terminal flower, the apical growth is terminated, the scarred apex of the main shoot is over-topped by a whorl of three lateral shoots." This must be noted as the prevailing characteristic of *Nerium odorum*. Although the usual arrangement of leaves is in a whorl of three, sometimes solitary leaves are seen. With regard to this disposition, the observation of Kerner is worthy of the notice of the student of Morphology. It runs thus-"If a whorl is composed of three leaves, and if the successive whorls displaced through one-sixth of the circumference as in oleander, six

* Vide " Laboratory Exercises in Botany", Philadelphia, 1895.

rectilinear series of leaves or orthostichies originate running parallel to one another down the stem."

The peculiarity in the situation of the stomata in this plant is worthy of special notice. The stomata are at the bottom of deep pits on the lower side of the leaf, and entrance to them is beset with extremely delicate hair-like structures. In its natural position the plant is much exposed to wetting by rain, mist, and dew, just when transpiration is an absolute necessity for it. But even when the leaves are covered on both sides with a layer of moisture, none can force its way into the hair-lined depressions which conceal the stomata, and consequently transpiration is not hindered even in the wettest season of the year (Kerner). Balfour* in describing the stomata says that they are furnished with cellular hair-like processes. It must be observed that the hair-like processes do not arise from the stomata, but from the epidermal layer of cells lining the deep pits in which the stomata are situated.

Mr. W.H. Gower† in his article on Green House Plants remarks that "notwithstanding the whole plant is poisonous, the splendid larva of the Oleander Sphinx moth (*Chaerocampa nerei*) however thrives well upon its poisonous leaves." In Rheede's Hortus Ind. Malab. Tablet I., there is an illustration of the double-flowered pink variety of *Nereum odorum*. There is also a butterfly, a caterpillar, and three figures of chrysalis. It is for an Entomologist to determine what the butterfly is.‡

Gower, in the article quoted above, makes a very useful suggestion for the purpose of securing showy and long-lasting flowers, which is worthy of notice. He says that "it sometimes happens that young shoots start out from the base of the flower stalk, which allowed to grow will entirely spoil the bloom." They must therefore be cut out as soon as seen.

The plant is of great antiquity. It was known to the ancient Greeks Romans. It was known to Pliny (Known as C. Plinius Secundus, Major) who flourished in the first century of the Christian era, and has been referred to Palladius Rutilius Taurus (Roman writer on husbandry who flourished in the third century of the Christian era), under the name of *Rhododaphne*, the rose-bay. This is what Pliny§ says regarding it — "Oleander called in Greeke Rhododendros which some

* P. 537 "Manual of Botany," 5th Ed. 1875

† Casell's "Popular Gardening," Vol. II, p. 316.

‡ Several species of the genus, *Euplæa* feed on the Oleander - L. de Nicéville.

§ "*Naturalis Historia*" translated by Dr. Holland, Vol. II, pp. 191-192, London name Rhododaphne and others Nerion hath not been so happie yet as to find so much as a name in Latine. A straunge and marvellous qualitie of this plant : the leaves are a very poison to all foure-footed beasts; and yet they serve man as a preservative and counterpoyson against serpents if they be taken in wine, with Rue among. Also sheepe and goats if they chance to drinke of the water wherein those leaves lay soaked, will (by reports) thereupon die." The plant is still to be seen, says Hehn, in Greece and Italy as oleander or rose-laurel, not only adorning gardens, but fringing the roads and dry beds of rivers with its fragrant rose-like blooms, and the faint brilliancy of its long evergreen leaves. (Nairne).

In describing the scent of the flowers, Balfour says that it perfumes the air for a distance around, and has the odour of almonds. The most penetrating odour is that of the double-pink variety, as compared with the odour of the rest of the variously coloured flowers, and resembles a mixture of the odour of bitter almond and otto of roses, I think.

POISONOUS PROPERTIES.

All the ancient and modern Hindu writers agree in ascribing poisonous properties to this plant. Two varieties are described in each of the following works — (1) Bhav-Prakash (Purva- Khanda, Part I, Shl. 79-80); (2) Madanpal Nighant. (Shl. 337- 338); (3) and (4) Dhanvantari and Raj-Nighant (p. 134, Anandashram Edition 1896, Poona). The following works describe five varieties — red, white, pink, yellow, and dark red — (1) Nighant-Ratnakar (Vol. I, p. 102, 1867, Bombay); (2) Nighant- Prakash (p. 52, Guna-Dosha, 1837, Bombay). All the works describe the several species as the "Killer of horses". In medicine, used externally, the root and bark are considered useful in skin- diseases and leprosy. On what authority Dalzell and Gibson call it a native of Arabia (Aden), it is not easy to understand. (See Supplement to the Bombay Flora, 1861, p. 52). English and Continental Botanists have from time to time dwelt on the poisonous effects the plant has on men and the lower animals. The stem, root, leaves, and flowers are all more or less very deadly. The following remarks of Lindley may be usefully quoted here — "Although little suspected oleander is a formidable poison. A decoction of its leaves forms a wash, employed in in the south of Europe to destroy cutaneous vermin; and its powdered wood and bark constitute at Nice the basis of an efficacious rat-poison.

A few years ago, a child died from having eaten one morning a quantity of oleander flowers; it was seized with violent colic, under which the child sank at the end of two days. In 1809, when the French troops were lying before Madrid, some of the soldiers went a marauding, every one bringing back such provisions as could be found. One soldier formed the unfortunate idea of cutting the branches of the oleander for spits and skewers for the meat when roasting. This tree is very common in Spain, where it attains considerable dimensions. The wood having been stripped of its bark, and brought in contact with the meat was productive of most direful consequences, for of twelve soldiers who at of the roast seven died, and the other five were dangerously ill. (Gard. Chronicle, 1844, p. 23). In like manner the root of *Nerium odorum* is found to be a poison in India." Why, then, I ask, is there no mention of this plant in the standard work of Alfred S. Taylor so sumptuously studied of all lawyers in Great Britain, Ireland, and India? I am writing from my knowledge of the Second Edition published in 1873. I have not seen the latest edition recently published. Equally strange is it to say that not even the erudite Dr. Casper, Professor of Forensic Medicine of the University of Berlin, makes any reference to oleander, although he refers to less virulent plants in the four volumes published by the New Sydenham Society, 1861-65. Stranger still, one of our latest writers on Forensic Medicine and Toxicology (1893, London), in the person of Professor Dixon Mann of Owen's College, Manchester, makes not the slightest reference to it, while Lauder Brunton and Schmiedeberg, Mr. Sohn, Dr. Dymock, and Mr. Greenish recognize poisonous alkaloids in *Nerium odorum*. This shows that even our first-rate writers on Toxicology are not in touch with the pharmacological knowledge of the day. This is not as it should be in the interests of human life and public safety. Lawyers defending criminals look up to works on Medical Jurisprudence and Toxicology for their sole information. I deem it my duty to point out that the works which Lawyers consider here and in England as their authorities are painfully wanting in placing before the public useful knowledge which has been handed down for hundreds of years as established facts — *facts beyond dispute*.

When O'Shaughnessy toiled in India, as Professor of Chemistry Maleria Medica in the Calcutta Medical College, in the first half of this century (*vide* Bengal Dispensatory, 1841, p. 445), the active principles causing dangerous symptoms had not been isolated, though he says many attempts were made to obtain the same. O'Shaughnessy, however, firmly believed that the poisonous principle was active and volatile. He even goes to the extent of saying that the vapour of the flowers in a close apartment will prove poisonous. The accuracy of this statement is to me doubtful. O'Shaughnessy says that the wood of this plant is used by some Eastern nations as the best material for gun-powder charcoal.

Coming now to discuss the results of modern researches, the following active principles may be mentioned as described by Sohn (Dictionary of the Active Principles of Plants, p. 65, 1894) —

1. *Oleandrine.* - Alkaloid? [Neriodorin, (Schmiedeberg); identical with one of Selmis's Ptomaines (*Finoshi*); statement of different observers concerning *Oleandrin, Neriin, Neriodorin, & c.,* are conflicting; see Leukowsky, *Rep. Chim.*, Appliq., III, 77; Schmiedeberg Archiv., Exp. Pathol., XVI, 151; Greenish, Pharm. J., Trans., 3rd Ser., XI., 873, and others]. Amorphous, yellow, resinous, bitter, poisonous. After heating to 240°C. it is no longer soluble in alcohol or water. Melting point above 56°C. with crystalline sublimate. Soluble in water, Betelli; alcohol, ether, chloroform and fatty oils.

Neriodorin scarcely soluble in water or ether, not soluble in benzene or petroleum ether. For further information regarding the precipitants, see Sohn.

2. *Neriin* has all the properties of Digitalin and possibly identical therewith.

3. *Nerianthin* bears similar resemblence to Digitalin.

4. *Rosaginin* — A Glucoside (*E. Pieszezck*). Amorphous, Archiv. Pharm., 1890, 228, 352; poisonous; action like Strychnine. Soluble in alcohol; not in water, ether, chloroform or petroleum ethers. For further tests, see Sohn.

The Presence of the Glucoside *Rosaginin* would seem to account for tetanic symptoms noticed in two cases reported in the *Indian Medical Gazette* of 1866. (1), In the case reported from the Calcutta Medical College Hospital the trunk limbs were found rigid, the jaw spasmodically closed & the pulse very feeble and slow, about 30.(2) Babu K.H. Acharji reports the second case in which tetanic symptoms were well marked. Babu D. Mukarji reports another case wherein similar symptoms were noticed. He pointedly observes that the action of the poisonous element resembles that of strychnia. He corroborates the statement made above regarding the extreme slowness of the pulse by adding that the marked difference between the effects of oleander and nux-vomica-poisoning consists in the slowness of the pulse. In nux-vomica-poisoning it is generally unaffected, becoming slightly quickened only during a fit; but in oleander- poisoning its preternatural slowness is a marked feature. (Dymock). This can be accounted for by the fact that this glucoside is not the only poisnous element in the imbibed potion; there are others, such as *Oleandrine* or *Neriodorin, Nerrin,* and *Nerianthin* which are distinct cardiac poisons. They stimulate the cardiac muscle to start with; they besides stimulate the Vagus Centre, as also the Capillaries. Lauder Brunton classes *Nerrin* along with *Digitalin, Digitaliin, Helleboreine, Scillain, Anriarin, Thevetine, Theversine, &c.* "The stimulation of the cardiac muscle is shown by increased energy of contraction, the rate of pulsation remaining the same or becoming slower. These are so-called Cardiac Poisons; with a large dose, the stage of stimulation is followed by one of peristaltic action, and final arrest in the systole. This stoppage of heart in systole occurs in frogs, but in higher animals the heart may stop in diastole.*

Dragendroff recognizes *Oleandrine* as the alkaloid found in the plant, but says he is not familiar with it, and refers the reader to the researches of Leukowsky[†]. Dymock, in referring to the researches of Leukowsky says that the latter recognizes in the leaves of oleander the presence of two alkaloids — namely, *Oleandrine* and *Pseudo-curarine*. Dymock further quotes the researches of Schmiedeberg, which, in view of the quotation from Sohn given above, referring to the difference of opinion as regards the nature of the true alkaloid, may well be repeated here. "Schmiedeberg (1883), who considers *Oleandrine* to be a glucoside, found in the leaves two other glucosides — *Neriine* and *Neriantine* : he considers *Neriine* to be identical with Digitaleine."[‡]

Greenish recognizes two bitter principles in bark, *Neriodorein* & *Neriodorin*, which, he says, closely allied non-nitrogenous substances,

* Pharmacology and Therapeutics, p. 276, 1885.

† Plant Analysis translated by Greenish, p. 204, 1884.

‡ Pharmacographia India, p. 401, Vol. II, 1891.

probably glucosides, both possessing the properties of powerful cardiac poisons. In the bark, he says, there are crystals of calcium oxalate (Year Book of Pharmacy, p. 154, 1881). Gribble classes the plant among irritants side by side with *Plumbago rosea* and *P. zeylanica.* T.N. Mukarji, Jaikisson Indraji, Sakharam Arjun, Lyon and Chevers note that the root is used to procure abortion. Lyon includes the plant among Cardiac Poisons. The post-mortem appearances which he quotes from Chevers bear out Lauder Brunton's remark, that in higher animals, such as men and quadrupeds, the heart may stop in diastole. For further information I would refer the reader to the observations of Chevers in his Medical Jurisprudence.

Veterinary-Major G.K. Rayment, A.V.D., wrote to me in March, 1896, from Rawalpindi that he was concerned as a witness in a case of poisoning camels by oleander. He subsequently at my request furnished me with a printed memo. of the symptoms. This memo. I am sorry to say, I am unable to find just now. Camels do not object to eating the leaves, but the leaves are invariably fatal to the poor beasts. Sir George Birdwood and Surgeon General Balfour have noted this fact before, in their respective works, namely, "The Bombay Products" and the "Cyclopaedia of India," 3 volumes.

Reprinted from
Journal Bombay Natural History Society
Vol. 11 : 606-623. 1896

PART XVIII.

CALOTROPIS GIGANTEA (*Br.*)
NATURAL ORDER — ASCLEPIADEÆ.
MARATHI — रूई (*Rui*).

A shrubby plant common in dry waste places, loose rocky soil, sandy ground, and old walls and hedges, throughout the Bombay Presidency. It avoids heavy clayey soil where its roots cannot easily penetrate. Milky throughout; glaucous or ash-coloured.

ROOT. — This is the most important part of the plant medicinally; often dividing and subdividing as far forward as its branches can penetrate; twisting slightly on itself before branching. The root-bark is pale or buff-coloured when fresh, $\frac{1}{8}$ to $\frac{1}{4}$ inch thick. Root-substance corky, soft, longitudinally wrinkled; taste bitter; milky juice acrid.

STEM. —Generally stout; when erect, 5-6 feet high, and 12-18 inches in girth; sometimes slightly reflexed; full of pith.

BRANCHES. — Often starting from the very base of the stem; thick, herbaceous all through; woolly, and covered over with a white mealy powder. This is well seen when the branches are young and fresh; it disappear as the plant grows old, or if it is roughly handled. As the branches grow old, the pith disappears. The branches are them hollow.

BARK. —The *Cortex or outer bark* is tomentose; ligth or pale buff-coloured; slightly corky and wrinkled longitudinally. The *Liber* or *inner bark* is green, and is mainly made up of strong, long, pale, silky, fine fibres of much commercial value. The Wood is soft, porous, light, cream-coloured or yellowish. Milky juice abundant.

LEAVES. - Exstipulate; opposite, rigid, often decussate; tomentose above cottony beneath the tomentum on the upper surface disappears on rough handling and from dried specimens. Hence it is that Hooker says the leaves are smooth above. Hooker also says the leaves are sessile, but the general character is best described by calling it sub-sessile, though at times the petiole is absent. Margin of the leaf quite entire. Generally 4" to 8" long, 1" to even 6" broad; cuneate, obovate. Base retuse, sub- cordate or even cordate; at times assuming what Prof. Bentley calls *Panduriform*. or fiddle shape as in Fiddle-dock (*Rumex pulcher*).

AXILS. — Bearded. This is corroborated by Wight in other words when he says : that "The leaves are often furnished at their insertion with glands or hairs in lieu of stipules."

PETIOLE. — Often very short, rigid, channelled on the ventral aspect, and bearing a beard of numerous, short, stiff-pointed brownish hairs where it merges into the midrib. Veins white and prominent on the dorsal aspect; alternate; sometimes in pairs, reflex at the margin. Venules transverse, thin. The midrib becomes gradually thinner and thinner on the under-surface, ultimately not more than a line thick as it reaches the apex, though at the petiole it may be as thick as a goose-quill.

INTERNODES. —Wide, often 3-4 inches long, and at times even longer where the stem grows quickly in suitable soil.

INFLORESCENCE. - Extra-axillary in sub-umbelliform or sub-racemose cymes. The umbels are cottony, says Hooker.

PEDUNCLE. — Alternate with the succeeding peduncle; rigid, solitary, many-flowered; with a white tomentum.

PEDICELS. — Subumbellate, longer than the flowers; pink, round.

BRACTS. -Oblong (Hooker). Hamilton says there are scarcely any.

FLOWERS. — Regular, pentamerous, hermaphrodite; tomentose; sometimes pure white, sometimes pinkish-purplish, the colour depending as the flower matures, and even for sometime after the flower has fully opened.

DISC. — Absent.

CALYX. — Very small, inferior, persistent; sepals five, hardly more than two lines in length, and one line in breadth, glandular at base ventrally; ovate, concave, acute; adhering to the base of the corolla.

ÆSTIVATION of the calyx quincuncial; that of the corolla valvate.

COROLLA. — Monopetalous, broadly campanulate. Petals 5, broad, smooth, spreading; $\frac{1}{2}$-2 inches in diameter, ovate lanceolate, acutre, revolute and twisted in age.

ANDROŒIUM —

STAMENS. — Five; adherent to the base of the corolla; alternate with the petals; with interposed linear glands, which when mature have brown stiff points.

FILAMENTS. — Connate, forming a very short fleshy tube or column round the Gynoecium. This structure is called the *Stylostegium* or, *Gynostegium*. This column bears on its dorsal aspect a series of five processes corresponding to the five filaments which is known as the *Staminal Corona*. This "Corona" is adnate to the anthers. Dr. Francis Hamilton thus describes this complex and quaint-looking structure, in his Commentary on Rheede's "Hortus Malabaricus" (Part II, Page 246, Transactions Linn. Soc., Vol. 14, 1825) — "Regular; formed of five compressed '*corpuseles*" which are attached longitudinally to the furrow on the dorsum of each of the filaments; which lie upon the anthers with an incurved tridentate apex; and which encircle the reproductive organs with an obtuse, spirally revolute base." I may repeat here Hooker's observations that "The coronal processes of the stamens radiate from the staminal column," and that "The Coronal scales are truncate and hairy."

ANTHERS. — Bilocular, erect, introrse; crowning the staminal column; connate; adnate by the connective to the sides of the stigma; tip often produced into an inflexed membrane (Hooker).

POLLEN. — Instead of being powdery is formed into ten *pollen masses* technically called *Pollinia*. These ten *Pollen-masses* consist of hundreds of pollen-cells (Kerner Von Marilaun). These pollen-masses are pendulous, lying in pairs outside the stigma in the form of yellow, flat, transparent, leaf-like bodies. Some describe them as pear-shaped; others as spatulate; DeCandolle says they are of the shape of a plough share. These differences in description are due to the structural developments of the pollen-masses, as examined by the respective observers. The figure in my plate of an enlarged pollen-mass, as seen under the microscope, and marked No. 3, is not materially, different from the description of any of the observers. Some more observations will be made further on, on this very interesting structure, known as the pollen-masses, in my general remarks. Kerner Von Marilaun says the *Pollinia* are in the form of shining, horny leaflets, and that they are attached to a clip- like body by ligulate strands.

GYNOŒIUM —

OVARY. — Of two distinct carpels, superior, and enclosed within the "Staminal column."

OVULES. — Numerous, anatropous, pendulous, multiseriate. Placentas on ventral suture; nerviform (Decaisne and Lamaout). Micropyle apical.

STYLES. — Two, short; subulate (*i.e.*, shaped like a cobbler's awl); closely adherent to each other in the stigma; but quite free below.

STIGMA. — Common to both the styles; dilated; depressed (Hooker); sulcated down the middle (Wight) ; five-lobed, forming an acute- angled pentagon; *the bases of the angles alternate with the anthers*; the angle itself each bearing a gland-like corpuscle terminating below in two processes which, about the time of the expansion of the flower, become firmly attached to the apices of the corresponding pollen-masses (Wight). Thus "Included between the anthers," as Hooker graphically describes, the stigma is short, broad, truncate, fleshy, almost cartilaginous; beautifully purple, and dark brown at the angles. The bases of the pentagon are fringed with a thin, whitish transparent fold of delicate structure.

FRUIT. — Consists of two distinct follicles, pointing opposite ways. Often one is abortive. The follicles are short, thick (Hooker); semi-ovate says Brandis. "Ovoid ventricose, green, herbaceous" (Dalz. and Gib.); glabrous, rugose, inflated. Between the outer and inner coats of the follicle there are several large fibrillar spaces.

SEEDS.- Quarter inch long, broadly ovate (Hooker); numerous, compressed, close-packed in a scaly manner, overlapping one another; hence termed by Wight "imbricate". This is well illustrated in the open follicle (Fig. 5). in the accompanying plate. Some Botanists describe the seed as marginate. Others say that the entire margin is slightly winged ovoid in form. The special feature of the seed is the comose tuft at its apex. When seeds lie close packed within the follicle, the comose tuft is placed away from the base of the follicle, so that as soon as the follicle bursts, the tuft expands & forces the seed out of the follicle. The comoe at times are very dense, silvery white & shining in the fresh condition, turning brownish on exposure, or by age. When the seeds ripen, the placenta shrivels. Thus detached, they lie ready to disperse. As soon as the follicle bursts at its ventral suture, the seeds fly about in the air, in profusion. It is quite a sight to see these scattered seeds in mid-air, the expanded comoe serving the purpose of a parachute. Dr. MacDonald happily terms them "seed-wanderers," in search of a new home. It is thus that they are seen invading loose soil, dilapidated buildings, and waste places, planting themselves in distant places, and propagating their species at times enormously. Thus it is that this plant does not require any human agency to propagate its species. I do not know of a single instance where this plant has had to be raised by sowing its seed.

ALBUMEN. — Thin, says Wight; so says also Gregg ("Text Book of Indian Botany," 1883).

EMBRYO. — Large; straight (Wight).

COTYLEDONS. — Flat (Hooker); foliaceous (Wight & DeCandolle).

RADICLE. — Short, inferior, says Hooker; superior, says Wight. I have had no time or opportunities to verify this. I leave this point to be settled by Morphologists (K.R.K.).

PLUMULE. — Inconspicuous (Wight and DeCandolle).

GENERAL REMARKS

SYNONYMS —

Asclepias gigantea — Hort. Beng. 20; Willd.; Roxb. Flora Indica, II, 30.

Asclepias gigantea — β. Lamarck, Encyclop. Meth., I, 280.

As regards bracts, Hamilton says "There are scarcely any." This is not accurate. They are deciduous, and have possibly escaped his attention. They are seen on almost every flower and flower stalk in the fresh state. Perhaps it would be proper to say that every flower, almost without exception, has its own bracteole (K.R.K.). Some Botanists call it an "*involucre*," but it is not that structure as we understand it in botanical parlance. There are various opinions expressed by able Botanists regarding the aestivation of this plant. There is no doubt about the quincuncial aestivation of the calyx. But although I have described the aestivation of the corolla as valvate, I feel bound to state that Wight describes it as *sub-valvate*, meaning "The very edges overlapping & *imbricate*". (See Wight's Illustrations of Botany, Vol. II, pp. 164-169). This observation of Wight's I feel fully bound to note as inaccurate, for "imbricate" means that one petal overlaps the adjacent one. This is not case in instance. The free ends of the petals fulluy fit into each other in a valvate manner. So, at the utmost, the aestivation could be termed *sub-valvate*, meaning that the petals are fitting closely to each other only at their margins; but, it must be understood that the petals are not on that account to be considered as "imbricate." With regard to the nature of the "staminal corona" and the pollen-masses, Professor Kerner Von Marilaun has made very important remarks in his "Natural History of Plants" (translated into English by Professor F.W. Oliver Vol. II, p. 258-259, 1895). They are well worth quoting here on account of the accurate description they contain, and are as follows — "Adnate to each of the five sides of the staminal column is a tumid bilocular anther with membranous wings running down its lateral margins. The wings are not appressed to the column, but are reflexed and stand out in pairs with the free margins of the two wings in each pair converging towards one another. This produces the same effect as if the anthers coating the central column were slit through longitudinally in front of every corner of the pentagon. Owing to the fact that the swollen part of each anther bears a curious excavated structure, it comes about that the pollen-producing portions of the anthers are nowhere visible externally save the membranous wings, or the five apparent slits." * * * "The hollow staminal appendages are full of honey, and each is embellished by a central process, shaped like a horn." I may add that the horn resembles the incurved horn of the Highland ram (K.R.K.). To continue the quotation, "At the back of every one of the five slits is a little clip-like organ from which proceed ligulate stands connecting it with the pollinia in the adjacent loculi of two different anthers, the pollen-mass in the left loculus of the anther to the right of the slit being thus associated with the pollen-mass in the right loculus of the anther to the left of the slit. The abundant honey in the staminal appendages emits a scent perceptible from afar, and attracts numberless insects to the flowers. The honey, being stored in a very superficial position, is accessible to insects with short proboscces," such as bees, wasps, and sand wasps. Ants are also among the commonest visitors to this plant in quest of honey. "During the time when the honey is most abundant, the flowers are either nodding or pendent, and they offer no convenient place for the insects to alight upon, or from which they can comfortably suck the honey. All parts of the flower are smooth and slippery, and the only way in which an insect can support its weight is by inserting its claws in the slits between the anthers. In endeavouring to take firm hold, the insect draws its claw from one end of the slit to the other, and so becomes attached to the clip-like organ at the back. When the insect's foot is withdrawn, the two pollinia adherent to the clip are dragged out of their niches. One of the claws on that foot is then seen to be wedged between the arms of the clip, whilst the two pollinia are suspended from it." The clip-like organs will be readily distinguished in Fig. 3 of the accompanying plate. It is coloured brown and is Y-shaped in the illustration. It would be interesting to follow still further the remarks of Professor Kerner Von Marilaun and trace the destination of the pollen- masses so torn away from their seat and to see how they are conveyed to the stigmas of other flowers. For, without doing so, this description of the pollen-masses must necessarily remain incomplete. "The question is," says the learned Professor, "where are these stigmas to be found? The pentagonal central column, surrounded by the five anthers, contains the ovary in its interior. The approaches to this organ lie through the so-called stigmatic chambers which are situated close beneath the truncate end of the column, and open outwards. The chambers are concealed in the slits, just as were the clip-like organs, and insects occasionally come across them as they move their claws about in the recesses. If the foot inserted by an insect has pollinia already attached to it, they are thus introduced into the slit in a new flower, and as the insect feels about for firm support, it thrusts the pollinia into the stigmatic chamber. When the foot is subsequently withdrawn, the ligatures attaching the pollinia to the clip are broken, and the pollinia are left in the chamber, whilst the actual clip maintains its grip of the claw." It may be stated here that the above quotation refers to the flowers of *Asclepias cornuti*, but the learned Professor remarks that other members of the *Asclepiadeae* have essentially the same mechanism, though differences in detail of course occur. I am persuaded to believe that the process of fertilization in the plant I am describing is not materially different from that depicted above in connection with *Asclepias cornuti*.

With regard to the substance which binds the clip-like organ to the stigma, it may be observed that Professor Kerner Von Marilaun terms it *Viscin.* It is a transparent structureless and colourless substance; it "Does not form droplets with water, nor does it dissolve in alcohol or olive-oil. It may be termed *Viscin*, from the similarity which it presents to the bird- lime obtained from the berries of the Mistletoe. (*Viscin*)." "It is very sticky, and on the slightest touch can be drawn out into delicate threads." "The sticky substance is probably a mucilage formed from the outer wall of the pollen- tetrad, or from the broken-down walls of the mother-cells." (pp. 101-102, op. cit.).

Professor Asa Gray very rightly remarks that the flowers of the entire order *Asclepiadeae* are rather too difficult for the beginner to understand. I would add that to the Amateur Botanists it is almost repulsive, if not a veritable *Pons asinorum.* Nevertheless to the truly faithful and earnest student of Vegetable Morphology, the complex structure of the reproductive organs of the *Asclepiadeae* is of immense, almost undying interest in common with that of similar organs among the *Orchideae*, *Aristolochiae*, and *Nepenthes* or Pitcher-plants. I may usefully add a passing remark here of Professor Kerner Von Marilaun's, which is one of very great importance in understanding the structural difference between the pollinia of the *Orchideæ* and those of the *Asclepiadeæ.* In the former, the pollinia are connected by a little knob called the *Corpusculum* "Which is soft and viscid; whereas in the latter there is a *clip* which is a hard dry implement with two arms capable of holding any small delicate object by gripping it. Secondly, the pollinia in the Orchids are clavate or of pasty consistency; in the Asclepiads, as already described, they are in the form of shining, horny leaflets" (Kerner Von Marilaun, p. 257, *op. cit.*).

There is a difference of opinion among Botanists with-regard to the flowering time of *Calotropis gigantea.* Rheede says it flowers thrice a year. But I have the high authority of Brandis & Gregg, & of Sanskrit writers such as Madan Pal (see Nighant, p. 31, Calcutta, 1886), coupled with my own experience, to be able to say that the plant flowers throughout the year, in suitable soil, wherever the roots of the plant can extend far wide & derive ample nourishment of the requisit kind. There are differences of opinion again as regards the odour of the flowers of this plant. The correct description of the odour, however, is to be found in Rheede's "Hortus Malabaricus" (Vol. II, p. 53), where he says that the odour of the flowers is by no means unpleasant; it is similar to that of lilies. I fully endorse this view, with this remark, that you have to smell the flowers just as they open, for the odour is evanescent. If that is not done, you will have to say with Brandis, that the flowers are *inodorous*. Brandis is one of our most accurate observers in botanical phenomena, and I have always held him in very high esteem. If, therefore, in agreeing with Rheede I contradict Brandis, it must be considered entirely due to my personal appreciation of the delicate odour of a fresh-blown flower. Moreover, I find that a Sanskrit writer, Bhav Misr. distinctly describes it as sweet-scented (Sanskrit, *Madhura*). [See Bhav Prakash, p. 101, Bombay, ed., 1890].

I am limited as to space, or else I would enter on the description of the Folklore of this plant, as prevalent in India among the Hindus. I would merely rest content with briefly saying that it is used on the third day of the bright fortnight of Bhadrapad —(Shukla-paksha, when the moon is waxing) during the worship of Sakhi and of Parvati, the wife of Shiva. This day is a general fast day for all Hindu females. The most rigid observers of this fast live on nothing but the leaf of *Calotropis gigantea* and a little sugarcane. The others abstain from their usual daily food and eat nothing but plantains, and *Chibud* (*Cucumis Melo*, Linn); and drink no water save what is contained in immature coconuts, and popularly, but mistakenly, called "Coconut Milk." The real coconut milk is the rich thick milk-like juice obtained from the matured "Kernel" or fresh "kopra" of the coconut. The leaf of this plant is besides used by Hindu ladies on the 5th day after accouchement in the worship of *Sati* and *Jiwati*. Surgeon-General Balfour of Madras, in his "Cyclopoedia of India" (Vol. I, p. 553), says that this plant is by the Hindus held sacred to Shiva. "Its buds," says he, "also form one of the five flowers on the darts with which Kama, The Indian God of Love, is supposed to pierce the hearts of mortals." In Bombay the Hindus offer the flowers to Hanuman, the Monkey-god.

The economic uses of the plant varied. The root, says Col. Drury, is used in the manufacture of gunpowder charcoal. Brandis says that in Kattywar and in the Dekkan young branches are used for that purpose. Charcoal for gunpowder is made in Kattywar says Captain Jacob (see Graham's Catalogue of the Bombay Plants, p. 121, 1839). Captain Twemlow mentions that the hill people about Mahableshwar obtain an intoxicating liquor called "Bar" from the plant (Graham, *op. cit.*). The wood is also similarly used says Surgeon-General Balfour. The silky down which caps the seeds is used by the natives on the Madras side in making a soft cotton-like thread. "It is susceptible of being spun into the finest yarn for cambric, and has been used for the manufacture of a light substitute for flannel, by Messrs. Thresher and Glennie of London. It is also being tried by Messrs. Cowan & Co. of Edinburgh, as a material for paper" (1885). I quite agree, however, with Mr. A. Smith, when he says that the silky hairs capping the seeds are too short to be used for any elaborate spinning purposes. The inner bark produces a strong, long, silky, tenacious fibre which is not affected by water. Its breaking weight is the greatest of all known vegetable fibres. A three-strand 3/8 inch rope can sustain 552 lbs. as against 407 lbs. of *Crotolaria juncea* (Sunn.), and 224 lbs. of coconut coir. This valuable fibre, as obtained from the inner bark, especially of the young branches, be it noted, is spoken of by Mr. A. Smith, as being capable of bearing a greater strain than Russian hemp. (See Lindley and Moore's "Treasury of Botany," p. 202, Part I, 1870). The fibre is employed for fishing lines, nets, gins, bowstrings, and even tiger-traps, in places where the plant grows abundantly and almost wildly. No wonder that the specific name of the plant is *gigantea*, when its fibre can be used successfully to entrap a tiger—one of the strongest and fiercest denizens of the Indian forests, and one whose existence is a terror to the surrounding population, the more especially when he is a "Man-eater!" To me, therefore, the specific name of "*gigantea*" appears to have been aptly chosen. I am not aware who the originator of the specific name is? I wish some better-informed student of Botany would help me and my readers to find this out.*

Since writing the above, I find that Mr. F. Gleadow, who is well known to the readers of this Journal, has published the following remarks in the *Indian Forester* for November 1897 – "As to quality, it is stated that the fibres of both species," (viz. : - *Calotropis gigantea* and *C. procera*) "are equally good. That is a statement that I have hitherto taken for truth, and have now to suffer certain qualms of conscience because I know the fibre of *C. gigantea* to be neglected in the Deccan, where it is common, and never suspected the possible reason, viz., its use lessness. However, that may be, I found the fibre of *C. gigantea* quite useless in

Firminger says that this plant is unquestionably a "Handsome flowering shrub," along with its congener *C. hamiltonii* (which, by the way, is no other than *C. procera*), and "nothing but their extreme commonness in the jungle and by the way-side excludes them from admission into the garden. ("Manual of Gardening," p. 532, H. St. John. Jackson's Edition, Calcutta, 1890).

Surgeon-General Balfour of Madras speaks of a *Manna* produced from this plant (*op. cit*), under the name of *Madar-ka- Shakar*. Dr. Dymock has the following remark to make a *propos* of this sugary exudation, in the Pharmacographia Indica (Vol. II, p. 430) — "According to Burhan, '*Ushr*' is a Persian name for all plants having a milky juice, and especially for the plant knwon in Hindustan as *Ak*. It would therefore seem that *Ushar* is not an Arabic word, as generally stated in the dictionaries, but of Arian (*Aryan*? — *K.R.K.*) origin, and perhaps connected with the Sanskrit '*Ush*' to burn." *** "The author of the Minhâj describes *Sakar-el-Ushar* as a gum which exudes from the inflorescence of the plant and gradually hardens. He remarks that people say that it is a dew which falls upon the plant and concretes like manna." I have examined hundreds of plants of Mudar (*Calotropis gigantea*), but I have failed to find the manna on any part thereof. When quoting the popular belief about the exuded

the Saharanpur District in the month of February." This is reproduced from the *Tropical Agriculturist*, Colombo, page 478, January, 1898. Contrast with this the following from the pen of M. Cathiravalo in the same Journal at page 472. This writer after trying the experiment of growing the plant on "Temple Bar," Sir Græme Elphinstone's Estate on Pallai, sent to England some cotton produced on the estate. Of course by cotton is meant the silky comose tuft on the top of the seeds. Mr. Cathiravalo says it was pronounced excellent. He adds : - "I have seen fishermen getting the fibre and making ropes out of it for their nets." At page 473 of the same Journal another writer signing himself B, writes thus about the fibre of *Calotropis gigantea* : I sent home a sample of this fibre some ten years ago unnamed. It was prepared by hand, regardless of cost, and the brokers classed it as the finest Rhea, value I think £ 36 or £ 38 per ton. I do not remember the percentage of fibre from the weight of stems cut, but it was rather higher than that from Rhea stems obtained at the same time : the latter were grown though under shade. *The Calotropis would doubtless be improved by cultivation.* (The Italics are mine - K.R.K.). Read in this connection the concluding portion of Mr. Gleadow's remarks - "As to cultivation I considered the matter twenty years ago, and came to the conclusion that it could not pay, and think so still, principally on the ground that the plant is of a straggling light demanding habit, and could probably not be grown dense enough to give any considerable yield. *but I made no experiments on the point, and that is the only reliable source of information.*" (The Italics are mine – K.R.K.)

gum "Which is said to fall upon the plant and concrete like manna," Dr. Dymock and his collaborateurs make no further remark than the following — "The best authorities describe its properties as similar to those of the plant, it would therefore seem to be nothing more than an exudation of the juices of the plant which naturally contain some sugar." Might I ask if *Calotropis gigantea* — the plant I am now describing, has any exudations, save the rich, milky juice when the plant is wounded, and the honey that pervades the staminal corona?

Surgeon General Balfour and several other Botanists say that the milky juice has been prepared like *caoutchouc* and *gutta-percha*, and yields 50 per cent. I can claim no personal knowledge in this direction.

The plant is known in Southern India as *Yercum*, and in Northern India as *Ak* or *Mudâr*. Note that *Ak* is a corruption of the Sanskrit Ark, meaning the Sun, and that *Mudâr* is a corruption of the Sanskrit *Mandâr*. In Bombay, the plant I am describing is invariably known among the Marathi- speaking classes as *Rui*. The Gujaratis call it *Akado*. In the Sanskrit works I have consulted, I find that Bhav-Prakash (*op. cit.*) recognizes two varieties of the plant namely, (1) the white, and (2) the red-flowered; so do Dhanvantariya Nighant, and Raj-Nighant (pp. 136-137, Anandashram Series, Poona, 1896).* Madan Pal in his Nighant (page 31, Calcutta, 1886) simply mentions two varieties of Mandar, but he does not specify the colours of the flowers of the two varieties. Mr. Woodrow describes two flowering varieties of *Calotropis gigantea* — purple and white. "The white variety", says this veteran and consummate practical Botanist, "is well worthy of a place in gardens, because some of the white flowers have as pure a colour as it is possible to find among flowers. White-flowered plants are met with rarely, over the great range of the common sort" ("Gardening in India," p. 384, Bombay, 1889). I remember having seen a similarly pure white-flowered plant in Mr. L.B. Joglekar's garden in Thana, two years ago, in a bush of pinkish-purple flowered Mudars, self-grown from the seed of a purple- flowered plant. To me this is an illustration of *lusus nature* among plants. Rheede has not failed to recognize the white and purple-flowered species.†

* As also Raghunath S. Lale, in his Guna Dosh Prakash, p. 143, Poona, 1892.

† It also prevails in the Sattara District among the purple-flowered plants.

I have already said that Graham, on the authority of Captain Twemlow, says, that "The hill people about Mahableshwar obtain an intoxicating liquor called *Bar* from the plant" (Catalogue of Bombay plants, p. 121, 1839). I may add here that Sir George Birdwood repeats this observation. (*Vide* his Vegetable Products of Bombay, p. 222, 2nd. Ed., 1865). Balfour observes that in Arabic authors on "Materia Medica," Mudar is said to have been known even to the Greeks.

POISONOUS PROPERTIES

In a previous communication to this Journal, describing the poisonous properties of *Gloriosa superba* (see Vol. VII, p. 492), I have already said that the leading Sanskrit writers recognize *Calotropis gigantea* among the nine secondary poisons (Upvisha) along with *Gloriosa superba* and seven others. (See also Nighanta Ratnakar, p. 255, Bombay, 1867). The leading Sanskrit writers are agreed that this plant is "heating," or "burning," in other words "acrid" and 'drastic'. Among the modern native writers of authority Rao Bahadur Kanny Lal Dey makes the following remarks about this plant — "The powdered *Root-bark* and inspissated juice are used extensively for their diaphoretic, emetic, alterative and purgative properties which have been known to the Indian practitioners for many centuries and regarded in some parts as *vegetable mercury*." Thirty to sixty grains of the powder are used for an emetic. "The author has lately found that a fluid extract of the *leaves* given in doses of one to five drops in intermittent fever, during intermission, generally cuts off the paroxysm more effectually than quinine. Poisonous in large doses" (Indigenous Drugs of India, p. 57, 2nd. Ed., Calcutta, 1896). Dr. Sakharam Arjun says "That the milky juice is a mild rubefacient, and is believed to produce abortion" ("Bombay Drugs," p. 85, 1879). The learned Doctor does not however specifically mention as to whether abortion is caused by the internal administration of the juice or by its topical application to the *os uteri*. Jaikisan Indraji says "That the milky juice is believed to procure abortion. It is a caustic, also an emetic and purgative." This writer also does not mention whether the abortive properties are displayed in internal administration of the juice or in topical application to the *os uteri*. More direct evidence is wanted on this score, as both Dr. Sakharam and

Mr. Jaikisan simply say *it is believed* to possess abortive properties. Personally, I have no knowledge of such use; I am therefore unable to express any opinion one way or the other. To come to the European and Anglo-Indian Pharmacologists who have studied this plant, O'Shaughnessy says that the powdered bark of the root, in doses of from half to one dram, "Proves emetic after an interval of from 20 minutes to an hour, generally causing much, nausea, and in about one case in every three inducing a cathartic operation. In doses of from 2 to 5 grains, taken every half hour, it proves nauseant, powerfully diaphoretic, and after several doses gently cathartic." An alcoholic extract of the milky juice, says O'Shaughnessy, "In doses of 10 grains acts as a powerful, but uncertain, cathartic and frequently causes violent vomiting." The dried bark of the root, says O'Shaughnessy, "Is of greyish-yellow colour; heavy and very peculiar smell, acrid, nauseous. It yields to water at 70°, 15 per 100 of gummy saccharine matter, and a peculiar extractive principle which has the singular property of gelatinizing as it is heated, then liquefying again and, as the solution cools, gelatinizing as before. This principle is termed *Mudarine* buy its discoverer, Dr. Duncan, at whose chemical and clinical experiments thereon, the Editor was present" — meaning O'Shaughnessy himself (Bengal Dispensatory, p. 453, Calcutta, 1841).

In passing, I may add that O'Shaughnessy feels satisfied that the powdered root is efficacious in incipient leprosy, for he says that the clinical experiments of Drs. Playfair Duncan and Royle leave no doubt of the bark of the root being *"really efficacious in incipient leprosy and in numerous diseases of the skin."** To enter into a critical examination of this very broad statement of three able exponents of the pharmacology of Indian plants, viewed in the light of recent clinical experience, would be beyond the scope of this paper. Dr. Duncan's discovery of the active principle of the root of *Asclepias gigantea*, Linn., as *Calotropis gigantea*, Br., was named in Dr. Duncan's days, marks an important epoch in the pharmoacological and clinical history of the plant I am describing. It is described by Dr. A. Duncan as an *"Extractive matter,"* in Phil. Mag., X, 465 (see Watt's "Dictionary of Chemistry," Vol. III, 1868). Dr. Duncan calls it *Mudarine*. A recent writer on the Active Principles of Plants, namely, Charles E. Sohn, drops the *e*, and calls the

* *N.B.* - The italics are mine – (K.R.K.)

extractive *Mudarin*. In doing so, he says, he is acting in conformity with the system adopted by the Chemical Society of London, in their journal of spelling all *alkaloids* with the termination *ine*; and all *glucosides* and *amaroids* (*i.e.*, Bitter Principles) with the termination *in*, without the final *e*. Starting with this clear understanding in the introductory and explanatory pages of his "Dictionary of the Active Principles of Plants," Mr. Sohn refers to Mudarin in his work at page 24 (ed. 1894, London) in the following terms — *Mudarin* is a bitter principle, *i.e.*, an *amaroid*. It is amorphus, yellow, soluble in alcohol and cold water, gelatinizing on warming. Insoluble in ether and turpentine. Professor Dragendroff of Russia (University of Dorpat) classes *Mudarin* among the bitter principles which have not as yet been shown to be glucosides, but which are sparingly soluble in ether; more freely in alcohol. (See Plant Analysis of Prof. Dragendroff, translated into English by H.G. Greenish, F.C.S., pages 175-171, 1884). Note also in connection with the remark which Prof. Dragendroff makes with regard to the sparingly soluble nature of *Mudarin* in ether as contrasted with Sohn's statement that *Mudarin* is insoluble in ether. I am not a practical chemist of any pretensions whatsoever, I am therefore not in a position to state which of the two eminent writers is accurate in his respective statement. I feel bound, however, to place before my readers the two statements, for the guidance of such of the practical chemists as are interested in the search of true chemical tests for *Mudarin*. The subject of practical organic chemistry is altogether beyond the scope of this paper, or, for the matter of that, beyond the scope of my present scientific studies.

But to return to the description of the poisonous properties of *Calotropis gigantea*, still further, from well-known authorities, Norman Chevers, – the pioneer of all writers on Indian Medical Jurisprudence, must first be quoted here in support of the poisonous properties of *Mudar*. At page 285 of his standard & unexcelled work on Medical Jurisprudence in India (Calcutta, 1870), Dr. Norman Chevers says thus-"In the Chapter on Infanticide in the last edition of this work, mention was made of the practice of murdering new-born female infants forcing the milky juice of the *Mudar* down their throats." In the "Indian Medical Gazette" of April and June, 1867, pp. 106 &148, Mr. McReddie, Medical Officer Hurdui, again drew attention this mode of infanticide, and suggested that an investigation should be made with a view to the detection of this poison by chemical analysis. He, at the same time, gave the results of some experiments on puppies. Mr. McReddie infers from them that Mudar juice is acrid, but that it is not, as it is supposed to be by the natives, a "*narcotic poison.*" Speaking for myself, I have not heard or read of the narcotic properties of any of the parts or products of *Calotropis gigantea.* It principally acts on the *Primae viae* so far as I know, producing at times violent emesis and drastic purgation, often far beyond calculation, and much to the detriment and even danger to life of the person in whom the powder of the root-bark is used as a cure for dysentery, and a vaunted *safe* substitute for the root of Ipecacuanha. I woudl rather use Ipecacuanha, now that we can have it in our dispensaries *sine emetine,* than expose a dysenteric patient to the risk of treatment of his complaint with such an uncertain substitute as the powdered root of *Calotropis gigantea.* I say this from experience, for eight years ago as Civil Surgeon of Thana, in medical charge of a large Depot Jail and of the Civil Police and Military Hospitals there, I obtained a large consignment of the powdered bark of *Mudar* and tried it at the earnest desire of my late lamented friend and former teacher of Materia Medica, Brigade-Surgeon Dymock, who was then our Principal Medical Storekeeper in Bombay. My experience lead to an utter discomfiture. Wherever I tried the powder of Mudar root- -bark, there was either violent emesis, or violent catharsis. The following year, in my official annual indent on Dr. Dymock, I asked not "for more." I have come to the conclusion that the vaunted position of Mudar as a substitute for Ipecacuanha is untenable, at any rate according to my clinical experience. But I stand corrected. *Apropos* of Dr. Norman Chevers, remark, cited above, that the milky juice of *Calotropis gigantea* is used for the purpose of murdering new-born, female infants, I find that Surgeon-General Balfour of Madras has the following remark regarding *Calotropis procera,* a near congener of *Calotropis gigantea* — The acrid juice of *C. procera,* says he, "is used by Rajputs to poison their infant daughters" ("Cyclopoedia of India," Vol. I, p. 553, 3rd ed., London, 1885). Brigade-Surgeon J.B. Lyon, C.I.E., who has been to me the chief inspirer of the papers on the Poisonous Plants of Bombay, which have so long occupied the pages of this journal, says thus regarding *Mudar* — "It may be mentioned that Mudar (a name, I may add, common to *Calotropis gigantea* and *Calotropis procera*) and tobacco are said to be used for purposes of infanticide" ("Medical Jurisprudence for India," p. 114, Calcutta, 1889). In Gribble's "Medical Jurisprudence" (Madras, 1885, p. 188) there are the following remarks regarding *Calotropis Hamiltonii*, which is no other than *Calotropis procera*, a congener of *Calotropis gigantea*, with respect to its abortifacient properties — *Calotropis Hamiltonii* (Madras native name Atrendo) "Is used both internally and externally. The milky juice is mixed with flour and given in a pill; a rag is then dipped in the juice and folded round a stick. About four and a-half inches are introduced *per vaginam*. This plan is useful in all stages of pregnancy, and there is not much danger to the mother or to the foetus, which latter may in fact be born alive if the step is taken at an advanced stage." I may observe that whether or not there be danger to the mother or to the foetus ultimately, the initial use of any part or product of this plant for the purposes of abortion has *Crime* at its bottom, within the meaning of the Indian Penal Code, and as such it is specially worthy of notice here.

In support of all that I have said above, I feel bound to refer to the researches of *C.J.H. Warden and L.A. Waddell. (Pharm. Journ.*, 3rd Series, XVI, 167-70). "In India, under the name of 'Madar,' two plants belonging to the N.O. ASCLEPIADACEAE, are known the *Calotropis gigantea* or *Asclepias gigantea*, and the *Calotropis procera* or *C. Hamiltonii*. The former is one of the commonest weeds throughout India, and is most abundant in the Lower Provinces and Eastern India; while the *C. procera*, which most closely resembles it, is found chiefly in the drier parts of North and Central India" ("Year Book of Pharmacy," 1886, p. 153).*

I cannot conclude this paper without referring to the most modern results of the three learned writers who have contributed to bring out the most standard work of the day in Anglo-Indian Pharmacology, under the name of the "Pharmacographia Indica." (See p. 428-437, Vol. II, 1891, Bombay). It is said, in this grandest work up to date on the Pharmacology of Indian plants, that *Calotropis* is not mentioned by Greek or Roman writers, but some Mahomedans give *Fijakiyns* as its Yunani name; this appears to be corruption word ηγαθεος "most holy" or "under divine protection," & was probably applied to

* I have found it in the Satara District since writing this paper (K.R.K.)

the plant by some Syrian physicians who instructed the Arabs in Greek medicine. The modern Persians call *C. procera* Khark and Darakht-i-Zahrnak or "poison-tree." So far back as 1813, Ainslie in his "Materia Medica" of Hindustran seems to have noted the purgative properties of both *Calotropis gigantea* and *Calotropis procera*. It is stated in the *Pharmacographia Indica* (Vol. II, p. 434) that the emetic properties of *Calotropis* were brought to the notice of the medical profession by Dr. Duncan who, as I have already noted before, was the first person to isolate the bitter principle of the plant and examine clinically, in the presence of O'Shaughnessy, the pharmacological properties thereof, so far back as 1829. (See Edin. Med. and Surg. Journ., XXXII., p. 65).

"Modern physiological research has shown that the juice applied to the skin acts as an irritant."

"In large doses *Calotropis* causes vomiting and purging, acting as an irritant emeto-cathartic." (Ph. Ind., p. 434).

This is the best description of the poisonous properties of the plant that could be given in the existing state of our knowledge regarding *Calotropis gigantea* and *Calotropis procera* as well. With these remarks I conclude this paper.

Reprinted from
Journal Bombay Natural History Society
Vol. 14 : 20-44. 1896

PART XIX.

DATURA FASTUOSA *Linn.*

NATURAL ORDER — SOLANACEAE.

MARATHI — धतूरा, (Dhatûrâ), धुत्रा, (Dhutrâ), धोतरा (Dhotarâ).

This is a herbaceous, coarse, rank-scented plant, with a colourless juice. Older parts of the plant, glabrous; younger and growing parts covered with evanescent whitish pubescence, otherwise known as "*silvery-down.*"

STEM. — It is irregularly zigzag; 1-2 ft. high, even 4 or 5 ft. sometimes; rather succulent from containing much pith.

BARK. — polished, greenish generally, with a puplish tinge most marked at the nodes; at times slightly rough either with minute scattered hairs, or when without hairs, showing scars wherever any hair existed in the early stage of the bark.

BRANCHES. — divaricate; marked with scars of fallen leaves.

LEAVES. — generally 3 to 6 inches long; triangular or ovate, but acuminate always, even if rounded.

LEAF-MARGIN — sometimes entire, oftener with a few large lobes; at times these large lobes have a few coarse short teeth. The leaves appear to the naked eye quite glabrous, but the magnifying glass often shows a very minute pubescence on the upper as well as the under side of the leaf. The leaves, as a rule, are glaucous-green above, and paler beneath. The MID-RIB is prominent below, with a dash of purple; so are its main branches. The VEINS are pellucid "sinuate and pinnate." The leaves, says Clarke, are "ovate entire or deeply toothed, glabrous." (Hooker's Fl. Br. Ind., Vol. IV. p. 242).

PETIOLE. — 1-2 inches long. STIPULES — absent.

FLOWERS — hermaphrodite, very large, solitary, erect from the node, but not axillary. Hooker remarks that the flower-buds in early stage are puberulous; so they are under a magnifying glass.

PEDUNCLES — short stout, purplish, solitary. BRACTS — absent.

CALYX — Tubular, free 3, inches long, sometimes a little longer; very minutely adpressed pubescent : circumsciss.

SEGMENTS - 5; often only 3 or 4; $\frac{1}{4}-\frac{1}{2}$ inch, wide; green, triangular, acuminate, acute. The lowermost portion $\frac{1}{4}-\frac{1}{2}$ inch, persistent and enlarged in fruit. Lobes at apex $\frac{1}{4}$ inch long, ovate- lanceolate, according to some writers.

COROLLA -tubular, funnel-shaped (otherwise called trumpet-shaped). TUBE — over 7 inches, generally, gradually widening upward. LIMB — recurved, 4 inches or more in diameter. LOBES — five; pubescent on the back; their plait or midrib terminating in a short linear or recurved acute tail. ÆSTIVATION — plicate or valvate in funnel-part of the corolla; and contorted at apex, i.e., spirally twisted in bud, as also after the flower is fully open.

ANDRŒCIUM

STAMENS — included; about as long as the corolla-tube.

FILAMENTS — filiform; adnate to the corolla-tube from base half-way up; free further on.

ANTHERS — linear; bi-lobed; each lobe nearly $\frac{3}{4}$ inch long. Dr. Trimen says they dehisce vertically, meaning of coarse-*longitudinally*.

CONNECTIVE — of the same length as the anther-lobes, filiform, less than half the thickness of the filament.

POLLEN — white.

GYNŒCIUM

OVARY — papillose; the papillae subsequently mature with progress of the pericarp of the fruit into sharp pointed short prickles. Doubt-less most of the papillae are abortive. The ovary is superior, 2-celled, or imperfectly 4-celled; seated on a yellowish DISK, 1-12 inch broad, and burying as much of the ovary in its substance.

OVULES — numerous, on prominent peltate white placentas.

PLACENTATION—axile.

STYLE — linear, over six inches in length; at times half an-inch longer than the stamens, at times half an-inch shorter.

STIGMA — shortly two-lobed, often conical, mostly capitate; 1/4 inch long, at times 1/3 inch.

FRUIT - A capsule sub-globose, 4-celled, 4-valved, nodding on a curved peduncle; supported at base on the reflected accrescent calyx. The pericarp is fleshy, green, covered with numerous scattered short straight sharp conoid prickles, the base of the cone resting on the pericarp.

DEHISCENCE — irregular (Trimen). Clarke describes it thus - "Capsule sub-indehiscent, or irregularly dehiscent near the apex."

SEEDS — very numerous, somewhat compressed, sub-reniform, close-packed; pale-brown when the fruit is mature and dry; albuminous. Trimen says that the seeds are "*nearly smooth*". The general testimony, however, is that the outer coat or the cuticle of the seed-is "*rugose*". It is so, not only when the seed is examined old and dry, but also when it is fresh. There is an excellent figure in Guy and Ferrier's Med. Jurisprudence (page 534, 5th Ed. 1881), of the rough honey-combed cuticle of the dry seed of *D. stramonium*, a congener of the plant I am describing, the seed of which latter, is not far different from that of *D. fastuosa*. A detailed description of the seed will be given further, where the Embryo is also described in detail. It is enough to say here that the embryo is curved round the central endosperm, that is to say the embryo is peripheric. In this the seed of *D. fastuosa* resembles in structure the seed of *D. stramonium*.

GENERAL REMARKS

The description given above is mostly based on that of Dr. Trimen's as published in Part III of his elaborate and exhaustive Flora of Ceylon, a work which it is much to be regretted, he did not live to complete. The first part of his Flora of Ceylon was published in 1893. The second part appeared in 1894, and the third in 1895. The climate of Ceylon, where Dr. Trimen worked zelously from 1879 as the Director of the beautiful Gardens of Peradeniya, proved fatal to him. He died on the 16th October, 1896 to the regret of the entire botanical world, without finishing his Flora of Ceylon. It was left to the venerable veteran botanist Sir Joseph Hooker to bring out the fourth and fifth parts of the work undertaken by Dr. Trimen. Dr. Trimen's Ceylon Flora must be considered an elaborate supplement to Sir Joseph Hooker's Standard Flora of British India. The latter is the conjoint work of several distinguished botanists, whereas the former is the work of a single hand, the product of the patient researches of a man possessed of a wide and marvellous capacity for original botanical work.

THE SYNONYMS of *Datura fastuosa* are numerous. I wish to devote a few lines to consider them with a view to show how the nomenclature of the Indian species of the Genus Datura has been determined by various European, American, and Indian botanists of note. It seems all the more necessary to do so, as it will be seen from Clarke's description of the N.O. Solanaceae (Hooker's Fl. Br. India, Vol. IV), that under Genus Datura, (pp. 242, 243), there are queries and doubtful varieties referred to, which are somewhat preplexing. It would be useful, therefore, to study page 720 of Fasciculus I of Sir Joseph Hooker's Index Kewensis (1893), where we find the following synonyms for *Datura fastuosa, Linn.* —

1. D. Ægyptica, *Vesl. & c.*
2. D. alba, *Nees.*
3. D. Bojeri, *Delile.*
4. D. dubia, *Rich.*
5. D. humilis, *Desf.*
6. D. hummatu, *Bernh.*
7. D. inoxia, *Mill.*
8. D. laevis, *Schkuhr.*
9. D. Metel, *Mill.*
10. D. muricata, *Link.*
11. D. Nilhummatu, *Dun.*
12. D. stramonium, *Thunb.*
13. D. Wagmanni, *Steud. Nom.* Ed. II; I. 484.

The following are the synonyms of Datura Metel, *Linn.*, as given in Hooker's Index Kewensis —

1. D. fruticosa, *Hornem, & c.*
2. D. guayaquilensis *H.K. & B.*
3. D. timoriensis, *Zipp.*

Under this head I may mention the synonyms of Datura Meteloides, D.C. *ex. Dun.*, especially as Emeritus Professor Marshall Woodrow of Poona College of Science mentions D. Meteloides in his work on Gardening in India (p. 401, 3rd Ed. 1899, Bombay). They are as follows —

1. D. Metel, Moc. and Sesse, ex Dun. D.C. Prod. xiii. 1.
2. D. Wrightii.

The following are the synonyms of Datura stramonium, *Linn.*, as given in Hooker's Index Kewensis —

1. D. Capensis, *Bernh.*
2. D. ferox, *Nees.*
3. D. loricata, *Sieb.*
4. D. lurida, *Salisb.*
5. D. parviflora, *Salisb.*
6. D. pseudo-stramonium, *Sieb.*
7. D. tatula, *Linn.*
8. D. Wallichii, *Dun.*, D.C.

Dr. Norman Chevers in his Medical Jurisprudence for India mentions a species of Datura under the name of D. ferox, on the authority of Dr. W. Palmer, as found in India. In Hooker's Index Kewensis, D. ferox is referred to China only. In Johnson's Gardener's

Dictionary also D. ferox is mentioned as a China plant, bearing white flowers. In Hooker's Index Kewensis the following synonyms are given for D. ferox —

1. D. Bertolinii, *Parl. & c.*

2. D. laevis, *Birtol. Fl.: Ital & c.*

To sump up, although Clarke in Hooker's Fl. Br. India describes D. fastuosa and D. Metel as separate species as originally named by Linnaeus, Dr. Trimen says that the two species are scarcely separable (Flor. Ceylon, p. 238-239, Pt. III, 1895). Fluckiger and Hanbury hold the same view. They say that D. alba, *Nees*, appears to be scarcely distinct from D. fastuosa, *Linn.* D. alba and D. stramonium according to these writers are different, as they say "the seeds of D. alba are very different in appearance from those of D. stramonium being of a light yellowish-brown, rather larger size, irregular in shape and somewhat shrivelled" (Pharmacographia, p. 462, 2nd Ed., 1879).

The Datura plant is mentioned in Amarkosh, one of the oldest Sanskrit Dictionaries extant. Its author's name is Amarsinha. Horace H. Wilson, a celebrated Sanskrit scholar, in the preface to the first edition of his Sanskrit-English Dictionary (1819), has it that Amarsinha flourished in the fifth century A.D.; the same view is held by another eminent Sanskrit scholar, Professor Monier Williams of Oxford. He says that the Amarkosha of Buddha Amarsinha belongs to a period not later than A.D. 500. (See "Indian Wisdom," Monier Williams, p. 171, Lond., 1875). Thus it will be seen that the Datura plant has been known in India for several centuries. There are many synonyms in Amarkosh for Datura. We are chiefly concerned with two of them. They are as follows —

1. उन्मत्त = (*Uṇmalta*). — This Sanskrit word is changed by European writers into *Hummatu* through the Tamil word *Umattai.* The same Sanskrit word is adopted by some European writers as *Nil-hummatu* through the Telugu word *Nalla-ummitte*, or *Nâllâ-umetta.*

2. मातुल = (*Mâtula*). - This Sanskrit word is changed by some European writers into *Metel.* I may here mention that the term *Meteloides*, which is used by some European writers, seems to be derived from the Sanskrit word मातुलुङ्ग (*Mâtulunga*). — It must be noted here that the three Sanskrit words mentioned above indicaṭe the white-flowered plant. For in Râja Nighant and in Bhâv-Prakâsh there is

the termकृष्ण धत्तूर(Krishna-Dhattura)=Black Datura. - It bears purple or violet flowers, single or double. Note, however, that the purple or violet colour is seen here and there on stems, nodes, and petioles, and *on the outside only*, of the single and double-flowered corollas. The inside of the corolla tube is pure white, or cream- coloured. There is no Datura flower which is absolutely as black as charcoal.

I find in Roxburgh's Flora Indica (p. 188, Carey's Calcutta Ed.), that Ramphius gives D. rubra as a synonym of D. fastuosa of Willdenow. Moreover I find in Dr. Norman Chevers' Med. Jurisprudence (p. 179, Calcutta Ed., 1870) that the purple species named in Sanskrit as *Krishna-Dhattura* is known in Bengal as *Kâlâ*, i.e., black Datura, or as *lâl*, i.e., red Datura. In Western India, especially in the Konkan from where I am writing this paper, the *kâlâ* Datura means the purple or violet variety. The red variety is unknown here. At least, I have not seen it, in the town and island of Bombay, in Salsett, in the Satara District (Dekkan), nor in the very heart of the Konkan-namey Ratnagiri District. I have been in the Ratnagiri District since 15th May 1898, examining the local flora. I have found no red flowered variety of Datura.

The third variety of Datura mentioned in old Sanskrit works such as Râjâ-Nighanta, Bhâva-Prakasha and others is named (*Râja Dhattura*)-the Royal Datura. Probably this included the double-flowered varieties in the olden days when the Sanskrit writers saw the plants actually growing before them. Under the common head (*Râja Dhattura*) are included Datura plants which bear white, bluish, purple or violet, yellow, or red flowers. I give the Sanskrit terms *seriatim* indicative of these colours, as follows —

1. सित (Sita) — White, single and double-flowered.

2. नील (Nila) — Blue or dark blue. Note that this Sanskrit word signifies blue or dark blue in relation to animals, plants, clouds, minerals, etc. From this appellation the Telugu name Nâllâ - Ummilte appears to be derived; the thence the name Nilhummatu of those European writers who studied the plant on the Malabar Coast where Telugu is spoken. This is probably the same as the following variety—

3. कृष्ण (Krishna) — Purple or violet-flowered variety. This is the same, I think, as Krishna-Dhattura-Datura fastuosa mentioned above, or perhaps, the double flowered variety of it.

4. पीत (Pîta) — Yellow-flowered variety, single or double-flowered or both. Both these varieties are found in India. I had them once in my garden in Thana. The single-flowered yellow varieties as named by European and American botanists and horticulturists are as follows —

D. humilis (pale yellow) ; D. Chlorantha and D. lutea (yellow).

5. लोहित (Lohita)-Red-coloured variety known in Bengal, according to Norman Chevers as लाल (Lâl) datura, i.e., either of the colour of iron-rust, or that of blood. It may also mean "of the colour of copper." I have not seen such a variety on this side of India. But Firminger of Calcutta has mentioned a double-flowered and a single-flowered variety also of the red or scarlet-flowered Datura named D. sanguinea (*See* p. 531, Manual of Gardening, 4th Ed., Calcutta, 1890). Firminger says that the flowers of this variety are of a deep red colour. The plant "thrives well in Ootacamund," he adds. Nay, he says, this, that he obtained from Ootacamund plants of this deep-red coloured variety of Datura for both the Calcutta Botanical Government Gardens and for his own private garden. But, says he, the plants soon perished seemingly unsuited to the climate of Calcutta. Note that the Calcutta Botanical Government Gardens, which I visited not long ago, are on alluvial soil on the banks of the Hooghly River in low-land regions; whereas the plants bearing blood red flowers were taken by Firminger for growth in Calcutta from the high-land plateau of Ootacamund, which is 6,000 ft. above sea- level. I do not know whether this red-flowered variety exists in the beautiful gardens of Bangalore. Dr. Cameron's elaborate catalogue of the plants growing or nursed in those gardens is just now not with me, or else I might have been able to say something more with regard to the prevalence of the red-flowered variety in Southern India. Practically the red-flowered variety is extinct in Western India. If it could not live in the lowlands of Bengal, it cannot possibly live, or if at any time it lived in the low-lands of the Konkan, it could not survive. In this connection I must add that in Roxburgh's Fl. Indica (*op. cit*, p. 188), D. rubra (*Rumph*) is mentioned as a synonym of D. fastuosa (*Linn.*). I gather from Hooker's Index Kewensis (Fasc. I, p. 720) that in America, among the Columbian plants, there is the single-flowered D.· coccinea and the single-flowered D. sanguinea of Peru. Then again, among the

double-flowered plants, there is the *D. fastuosa-rubra* of South America. The following note on D. sanguinea (Ruiz. et. Pav.) by M. Berthold Seeman, Naturalist, H. M.'s Herald, may perhaps interest my readers as an illustration of popular faith in the plant-a mere superstition-among the American Indians of Darien and of Choco. These superstitious people of tropical America prepare from the seeds of the plant a decoction, which is given to children to produce a state of excitement in which they are supposed to possess the power of discovering gold. In any place where the unhappy children happen to fall down, digging is commenced; and as the soil nearly everywhere abounds with gold- dust, an amount of more or less value is obtained (p. 170, Vol. XI., Pharm. Journal 1852). This passage I find particularly worth quoting, as in India the administration of datura seeds to unsuspecting victims is not for collecting gold likely to be found in Indian soil-dust, but for searching the pockets of travellers and shop-keepers and depriving them of the gold and silver that may be with or about them, after they are well stupified with Datura seeds administered in various articles of food and drink. This will be amply evident in my remarks to follow under the head, of "Poisonous Properties."

I wish to dwell for a moment in naming the DOUBLE-FLOWERED varieties of the genus Datura, as in my experience. I have found them of great garden-beauty, having grown them in my garden, in Thana and Ratnagiri. The double-flowered varieties have been named (by writers well worthy of recognition) as follows —

I. WHITE (a) Datura cornigera — florepleno, (b) Datura Knightii.

II. PURPLE. — Datura fastuosa - florepleno.

III. YELLOW. — Datura chlorantha-florepleno.

IV. PURPLE - WHITE. — Datura fastuosa- rubra (S. America).

Synonym — D. Wagmanii (Hooker's Ind. Kew.).

It must be noted here that in the double-flowers Datura of all colours the anther-bearing stamens are changed into petals or perianth either antherless, or bearing anther-lobes more or less modified, on the extreme margin of the inner corolla-tube. "Indeed", says Kerner, "there are grounds for believing that all petals are originally modified from stamens." Be it noted here that double-flowers remain on the plant on which they grow, two or three days longer than the single-flowers.

I now proceed to say a few words about the SCENT of the flowers. Opinions on this point vary; there is the old saying— "*Quot homines, tot sententice*-as many men, so many opinions. This is the same as saying in the gustatory line, *de gustibus non disputandum.*" Lindley says that the flowers of Datura are sweet-scented, especially at night. Note that the flowers of Datura of all kinds open about sunset, or just after, and close about or soon after sunrise. Bishop Hebr's lines are well worth quoting here —

"The broad Datura bears her breast
Of fragrant scent, a virgin white,
A pearl amidst the realms of night."

These lines are somewhat differently worded in Rev. Mr. Nairne's "Flowering Plants of Western India" (p. 209, 1894), but the fragrance of the flowers is referred to. Mr. Donald McDonald says the Datura shrubs produce amidst a mass of elegant foliage large and fragrant trumpet-shaped flowers. "Those flowers that are lasting may be taken under glass to impart a delicate yet powerful scent throughout the green-house."*

As regards the odour of the leaves of Datura the general testimony is that they are rank, smoky. The odour is characteristically offensive; herbivorous animals shrink from it , says Kerner. Lauder Brunton says that the leaves have a heavy odour, which is strongest while they are drying, and of a mawkish faintly bitter nauseous taste. Over four hundred years ago John Gerarde, of London, made the following remarks on the Thorn-apple plant in an elaborate and wonderfully accurate work entitled "The Herball" — "The flowers are of strong ponticke savour offending the head when smelled unto. * * * The herbe itself is of a strong savour and doth stuffe the head and causeth drowsinesse." (P. 347). In Sowerby's later work first published about the middle of the nineteenth century, and entitled "British Poisonous Plants," it is said that the leaves of D. stramonium have a slightly foetid odour, but the flowers are sweet-scented though producing stupor if their exhalations are breathed for any length of time. (P. 29, 2nd Ed., 1861, London). "The whole plant smells of bean meal." (P. 134, Loudon's Encyclopaedia of Plants, 1829, London). The flowers, says Loudon, have an agreeable odour

* *See* p. 37, 1895, "Sweet-Scented Flowers"

first, but if smelt long they become less agreeable, and are narcotic. Whether the odour of this flower or that of the several species or varieties of the Datura genus be considered sweet or otherwise, fragrant or offensive, depends entirely on individual capacity to appreciate the differences or even the degrees of scents. Mr. J.Ch. Sawer, F.L.S., very wisely remarks, that "odours are differently appreciated by different people, and what pleases one person may have a reverse effect on another; thus the strong odour of *Tagetes patula* (French marigold) and *Tagetes erecta* is not unpleasant to some, while others consider it very objectionable." (Odorographia; Introduction, p. xvi. 1892, London). The same remark may be applied to the Datura plants.

Apart from the fact that the Datura plants are rank weeds, growing amidst hedges, fields and dunghills, where no human hand has sown them, often prevailing where they are not wanted, it is unquestionable, that the Thorn-apple is artistically speaking, a plant of great beauty, and quite a garden-ornament. Dr. Christison of Edinburgh remarks in his work on Poisons (1845), that the Thorn-apple in his day had become quite an ornament of Edinburgh Gardens. Firminger of Calcutta, on the other hand, is quite of a different opinion, although he is an able writer on Gardening. He would destroy the Thorn-apple plants after their flowers cease to bloom. For, he says, "they take up much room and look unsightly." Evidently Firminger has not the artistic eye or talent of Ruskin, or even of Kerner. The latter observes, that in the Datura plant the various forms, and the distribution of the green leaves, young and old, on the surface of the stem is very characteristic. Their position and form afford much room for observation. The unequal size of adjoining leaves on the same branch or stem is quite an artistic phenomenon. Looking down upon a horizontally projecting branch of the Datura plant, the larger and smaller leaves will be found arranged in quite a peculiar and striking manner. The smaller leaves are seen in the gaps between the larger ones. This mosaid like fitting together of larger and smaller leaves appears to be combined with the want of symmetry of the leaf-base most marked in the old long-stalked leaf.* It requires an artistic eye to appreciate these observations. It is not always that

* *See* Kerner's Nat. Hist. of Plants, Oliver's English Version, Vol. I, p. 422, 1894.

a gardener's eye is artistic. It appears from London (*op. cit*), that the Stramonium plant was known to the Greeks as the mad-apple. It is not mentioned by Professor Daubeny in his Oxford Lectures on the Trees and Shrubs of the Ancient Greeks and Romans.

POISONOUS PROPERTIES

All the species or their varieties hitherto known of the Datura genus are decidedly poisonous. The purple or violet coloured variety is more deadly. All writers, European, American, or old Indian, are agreed that every part of the plant is deleterious to human life. The seeds are the most poisonous of all parts of the plant, wherever it grows. A detailed description of the seed, therefore, may not be out of place here. I give it on the authority of Dr. William Palmer (Norman Chever's Med. Jurisprudence for India, pp. 184 - 185, Calcutta Ed., 1870). It runs thus—The seed is almost kidney-shaped; its outline, angular; its size is rather more than a quarter of an inch long, and rather less in width; its colour greenish-brown when fresh, changing to yellow, I may add deep brown, when dry; it is attached to the placenta by a large white fleshy mass, which separates easily, leaving a deep furrow along half the length of the concave border of the seed; the outer surface of the seed is scabrous, almost reticulate, except on the two compressed sides, where it has become almost glaucous from pressure of the neighbouring seeds; the convex border of the seed is thick and bulged, with a longitudinal depression between the bulgings caused by the compression of the two sides. When the seed is divided into two, by cutting with a knife placed in the furrow on the convex border, the testa is seen irregular and angular in outline, and the embryo is curved and twisted in a fleshy albumen.

The active principle of the plant is an alkaloid once known as Daturine. The seed contains it in larger proportions than any other part of the plant weight for weight. The alkaloid was also known at one time as Daturia. Sohn says that commercial Daturine is frequently a mixture of Hyoscyamine and Atropine or the former solely. *Datura stramonium,* he says, also contains Stramonine which is an alkaloid like Hyoscyamine and Atropine, but it is not bitter. Hyoscyamine has a sharp and disagreeable odour;

Atropine has a disagreeable metallic taste.* Erhardt and Poehl dispute the identity of Atropine and Daturine, says Sohn. Professor Dragendorff[+] says that "according to the more recent researches of Ladenburg, henbane contains two alkaloids, one of which Hyoscyamine, is isomeric with Atropine, and identical with Daturine and Duboisine." Ladenburg distinguishes Hyoscyamine from Atropine by the melting-points of the alkaloids, and their gold- salts. Professor Schmiedeberg of the University of Strassburg says[++] that Atropine occurs in Daturine of the Thorn-apple; Hyoscyamine, which is isomeric with Atropine is also said to be contained in the Thorn-apple. But he doubts the identity of Duboisine with Hyoscyamine[$]. Dymock and his collaborateurs who have carefully examined the plant, say that Prof. E. Schmidt and Mr. Schute have found, as the result of their researches, that the seeds of *D. stramonium* contain much Hyoscyamine, with small quantities of Atropine and Hyoscine (*Apoth. Ztg.*, 1890, 511). Stramonine is not mentioned along with these. But in Dymock's Ph. Indica it is stated that M. Gerard has prepared a new fat acid, *Daturic acid*, from the seeds, which yield 25 per cent. Oil when extracted by ether. Purified with petroleum, this oil is of a peculiar greenish-yellow colour. M. Gerard places Daturic acid between Palmitic and Stearic acids. They have analogous properties. Daturic acid crystallizes by cold from 85 per cent. alcohol giving groups of fine needles. It is fairly soluble in cold alcohol and very soluble in ether and benzene. I have no comments to offer on the quotations I have given above from celebrated pharmacological investigators; but I have yet to place before my readers the most recent opinions expressed by Dr. Murrell[||] as regards the nature of the active principles found in the different species of the datura plant and some of its congeners such as Belladonna, Hyoscyamus and Duboisia myoporoides. Dr. Murrell says that according to the old classification the active principles were as follows —

1. Belladonna contained Atropine.
2. Hyoscyamus—Hyosciamine and Hyoscine.
3. Stramonium—Daturine.

* See p. 14, Sohn's Dictionary of the Active Principles of Plants, 1894, London

+ Plant Analysis - English Translation by Greenish, p. 60, 1884, London

++ Elements of Pharmacology, Dixon's English Translation. Edinburgh, 1887

$ Pharmacographia Indica, p. 588, vol. II, 1891, Bombay by Dymock, Warden, and Hooper

| | A Manual of Pharmacology and Therapeutics, Pages 430-452, London, 1896.

"Ladenburg," adds Dr. Murrell, "has *re-investigated* the matter and says there are only three natural mydriatic alkaloids,* They are as follows —

1. *Atropine* — which occurs in Atropa belladonna, and in Datura stramonium (I may add in the Indian Datura, varieties and species of all Datura plants, K.R.K.).

2. *Hyoscyamine* — which occurs in Belladonna, Datura, Hyoscyamus and Duboisia myoporoides.

3. *Hyoscine* — which occurs in Hyoscyamus.

Dyboisine is identical with Hyoscyamine; and Daturine is a mixture of Atropine and Hyoscyamine.

N.B. — I am all at sea here, for Sohn, whom I have quoted above, says that the Datura stramonium contains Atropine, Byoscyamine, Hyoscine, and Stramonine. The reader may accept the views of any of the pharmacologists I have cited above.

The term "Atropa," says Murrell, "is derived from *Atropos* (Gr.), one of the evil destinies, whose mission in life was to destroy life, and it is supposed to be indicative of the fate of those who came under its influence." These "evil destinies"—so named by Dr. Murrell-were known as the *Parcae* or the *Fates* among the ancient Romans. They were three—(1) *Clotho*, or the spinning Fate; (2) *Lachesis*, or the Fate assigning to man his fate; (3) *Atropos*, or the Fate that cannot be avoided. (See Dr. William Smith's Classical Dictionary, p. 455, London, 1868). Reader mine, pray pardon this classical intrusion! It is at times refreshing to fly to old classical literature, in considering the hard facts of this our scientific age.

Dr. Murrell says that Atropine Hyoscyamine and Hyoscine are isomeric, each answerable to the formula $C^{17}H^{23}NO^{3}$. They can all three be resolved thus —

1. *Atropine* yields tropic acid and tropine (base);

2. *Hyoscyamine* yields the same, *i.e.*, tropic acid, and tropine (base);

3. *Hyoscine* yields tropic acid, and pseudo-tropine.

Note here, that, in chemical parlance, Atropine is a compound of a base called Tropine, and Tropic acid. Ladenburg calls the compound Tropine.

I pass on now to consider the action of the alkaloids hitherto recognized in the Datura plants and its congeners Atropa belladonna, Hyoscyamus niger and Dubosia myoporoides.

* *N.B.* - Dr. Murell does not give the date of the re-investigation made by Ladenburg, whose name has been mentioned in my quotations in the foregoing remarks - *K.R.K.*

I. ATROPINE — The medicinal dose prescribed in the British Pharmacopoeia is 1/200 to 1/100 grain. It is in comparison with this dose that the following remarks must be supposed to be made.

(1). *General action on man* — A full dose (B.P.) produces great dryness of the tongue and roof of the mouth, extending down to the pharynx and larynx, giving rise to frequency and difficulty in swallowing, and exciting a hard, dry cough. The face becomes flushed, the eyes are bright and injected, the pupils are dilated, the sight is dim and hazy, whilst the power of accommodation for distant objects is lost. There is mental disturbance often amounting to decided delirium, the delusions, as a rule, being of a pleasant nature. The patient is extremely restless, and cannot be kept quiet. The skin is dry and a rash appears closely resembling that of scarlet fever. (Murrell).

(2) *General action on the lower animals* — Pigeons and rabbits are; according to Murrell, almost insusceptible to the action of Atropine; so are horses and donkeys. "As a rule," says Murrell, "vegetable feeders do not respond readily to its action, the most pronounced effect being observed in the class of flesh eating animals."

(3) *Action on the Heart and Circulatory system* — In most animals there is an increase in the frequency of the pulse. "The first effect in man," says Murrell, "is to increase the frequency, fulness, and force of the pulse to the extent of fifty to sixty beats in the minute." Atropine paralyses the pneumogastric nerve. This may be due to an action on the trunk of the Pneumogastric nerve (otherwise briefly called Vagus), of on its peripheral terminations, or on the intra-cardiac branches which terminate in the intra-cardiac ganglia. Lauder Brunton puts it briefly that Atropine paralyses the efferent Vagus-ends in the cardiac ganglia. Moreover, says Murrell, "Atropine stimulates the vaso-motor centre, and so contracts the blood-vessels, and heightens the arterial pressure."

(4) *Action on the Respiratory System.* — A large dose of Atropine accelerates respiration. This is due to the stimulation of the respiratory centre powerfully . In consequence, the chest- movements become deeper and more frequent. This effect is independent of blood pressure. (Murrell). A large dose of Belladonna, says Dr. Ringer, will sometimes induce dryness of the *Schneiderian* membrane, which

though nasal, forms a part of the respiratory, and not of the olfactory tract of the nose.

(5) *Action on the Nervous System.* — The delirium following Atropine-poisoning shows that Atropine acts on the cerebral cortex. Hence the symptoms noted above and others not noted,—*viz.* — "exhilaration of the mental functions, giddiness, restlessness, and automatic chorea-like movements. There is usually loud, disconnected talking delirium and raving. The delusions are of a pleasing nature, and weeping or lamentation is rare." (Murrell). With regard to the Kind of delirium, Lauder Brunton says, that it is characterized by being very active and busy; the patient always wants to be doing something. At the same time, says Lauder Brunton, "this very business is accompanied by a great deal of languor and disinclination to move, because the peripheral ends of the nerves are weakened, and so there is difficulty in the way of the excited centers causing any movements in the muscles. It is only in very large doses that Atropine will paralyse the motor nerves completely, or at least the motor nerves going to the voluntary muscles."

(6) *Action on the Muscles* — Read with the foregoing views of Lauder Brunton quoted above, the Murrell observes thus—"The voluntary muscles are not affected." "The unsteady gait," says Murrell, "often noticed in man, is due to an action on the (spinal) cord or on the motor nerves, and not on the muscles." Murrell further observes that Atropine increases the contractile power of involuntary muscular fibre, (such for instance as is found in the heart and in the intestines of man - K.R.K.). "But," says Murrell, "it has been maintained that the increase in the persistaltic movements of the intestines is due to depression of the inhibitory branches of the splanchnics."

(7) *Action on the Glandular System.* — As noted above, "one of the earliest and most notable effects of Atropine is dryness of the mouth, from suppression of the secretions of the mucous and salivary glands." (Murrell) This, says Lauder Brunton, is due to the paralysis of the salivary nerves. *Apropos* of this, Murrell makes the following remarks — "According to Heidenhain's hypothesis with regard to the salivary glands, there are two kinds of secretory fibres, one, the *secretory*, the other, *trophic*, causing an increasing in solubility in the stored-up gland-substance. On the assumption of the different

kinds of secretory fibres there is ground for supposing there is a *third* variety — *anabolic fibres* — causing the formation of fresh substance by the cells. After an injection of sulphate of Atropine there is no increase either in the percentage of salts oin the sympathetic saliva produced by stimulation of the *chorda tympani*, as there would be if the trophic fibres of the *chorda* escaped paralysis. It would seem that atropine paralyses the trophic as well as the secretory fibres of the *chorda tympani.** Atropine checks the sweat secretion, says Murrell, by paralysing the efferent sweat-fibres which accompany the vaso-motor fibres. Lauder Brunton corroborates these remarks, for he says, in small doses Atropine will paralyse the efferent nerves which end either in voluntary muscular fibres, in glands or in ganglia. Atropine paralyses the lacteal nerve-terminations in the mammary gland of the human female. The secretion of milk is consequently arrested. Atropine similarly arrest secretion of the pancreas. The action of the liver is also affected. The quantity of bile is lessened. Murrell says that the effect on the urinary system is somewhat doubtful.

(8). *The Rash on the Human Skin.* — Dr. John Harley affirms that generally it is nothing more than of a mere temporary kind; "but it rare cases, and in persons who are liable to vascular irritation of the skin the redness remains, and its disappearance is attended by slight roughness and desquamation." He mentions two cases, in one of which "the patient was scarlet from head to foot," and another in which after the fourth dose, there was a scarlations tint of the skin." Dr. Gillespie has met with a case in which after injection of a small quantity of extract of belladonna into the urethra of a patient, the patient became "as red as a lobster" in less than five minutes (Murrell)†. Lauder Brunton also notes that a red rash appears on the skin like that of scarlatina‡. Schmiedeberg says, "the redness of the skin resembling scarlet fever *so often observed*, especially in the upper part of the body, and the similar colour with turgescence of the featuers, are probably connected with increase of the frequency of the pulse, with the increase of blood pressure caused thereby, and with the simultaneous dilatation of the vessels of the skin."§

* Murrell, *op.cit.* p. 437 † Murrell, *op.cit.* p. 438.

‡ Text book of Pharmacology, p. 904, 2nd Ed., 1885, London

English Translation, pp. 53-54 (Edinburgh, 1887), by Dixon (Sydney University, N.S.W.)

I have said above that Atropine checks perspiration and causes dryness of skin. But in one case of poisoning, says Schmiedeberg, "the burning hot skin was here and there covered with perspiration (Gerson). The perspiration that comes in this manner has the same origin as that in the death-agony." Such sweats, cold and clammy, are the precursors of death. In contrast with this case, it may be noted that Lauder Brunton says that the temperature of the human body is increased by small doses of Atropine and lessened by large ones.

(9) *Action on the Eye.* — When Atropine is applied topically to only one eye, it dilates the pupil of that eye alone, and not of the other eye. When Atropine acts indirectly, *i.e.*, through the circulation of blood, both the eyes have dilated pupils; the eye becomes bright, dry, and injected both by topical application and through circulation. "The power of accommodation is lost, and after large doses intraocular accommodation is lost. When Atropine is applied locally so as to affect the pupil of one eye only, the large amount of light entering through the dilated pupil produces contraction of the pupil of the other eye." (Murrell.) The pupil is normally under the control of the two antagonistic mechanisms—(a) The *contracting* mechanism, reflex in nature, of which the *Third Nerve* acts as an efferent, and the *Optic Nerve* as the affernt tract. (b) The *dilating* mechanism, tonic in nature, of which the cervical sympathetic nerve is the *efferent* channel. Murrell says that when the *Third Nerve* or the *Optic Nerve* is cut, the pupil dilates from the action of the *Sympathetic.* When the *Sympathetic* is cut, the tonic dilating influence ceases, and the pupil contracts. On stimulating the *Third Nerve*, or the *Optic Nerve*, the pupil contracts. On stimulating the *Sympathetic*, the pupil dilates. The dilatation of the pupil produced by the local application of Atropine might, at first sight, be attributed to paralysis of the *Third Nerve.* This view is untenable, says Murrell, for when the *Third Nerve* is cut and the pupil dilates under the influence of the Sympathetic nerve fibres, the application of Atropine, still further, dilates the pupil. From this it follows that Atropine says Murrell, exerts an action on some local mechanism. This mechanism, says Murrell, is probably situated in the *Iris*, or in the *Choroid*, where ganglionic cells are abundant. The paralysis of accommodation is due to the paralyzing action of Atropine on the Oculo-motor

Nerve terminal ends. Note that Ringer observes thus — "A large dose of Atropine will sometimes induce dryness of the Schneiderian membrane. This does not necessarily mean loss of the sense of smell." It must be observed here, that the foregoning observations are mainly intended for such of my readers as have a fair knowledge of Human Anatomy and Physiology.

Murrell observes that of all the tropine alkaloids, Atropine is the slowest in inducing its effect on the eye, but it lasts a long time, even many days. Homatropine induces its effects rapidly, but they disappear in a few hours. Hyoscyamine, in this respect, occupies an intermediate position.

This leads us on to consider the special actions of Hyoscyamine, which is one of the alkaloids found in Datura plants.

II. *Hyoscyamine* or Hyoscyamia is found in several Datura plants. Murrell says "it is *ismomeric* with Atropine and Hyoscine. It is *identical* with Duboisine, and mixed with Atropine constitutes what is known as Daturine. It may be split into tropine and tropic acid." When pure it is in snow-white minute crystals, soluble both in spirit and water. A substance is sold under the name of *Amorphous hyoscyamine*, which is mixture of Hyoscyamine and Hyoscine. It is a dark brown substance looking like an extract, and has a strong disagreeable odour.

The following observations are curtailed from Murrell, as regards the active principle Hyoscyamine. Hyoscyamine dilates the pupils (but moderately, K.R.K.); dries the mouth, and arrests secretions; flushes the face and produces a rash on the skin. It gives rise to a drunken gait, and excites delirium and hallucinations, but more frequently acts as a narcotic, inducing comatose sleep-the very sleep of death. As a rule, says Murrell, the raging delirium is not present but there is a desire for rest and sleep, probably due to the *Hyoscine* it contains. As noted above, it is a less powerful mydriatic than atropine- midway between atropine and homatropine. Schmiedeberg quotes the following remark from V. Schroff—"Hyoscyamine acts on the brain somewhat differently from Atropine. In men, after the administration of the *Amorphous modification*, the raging delirium is as a rule not present, but even after small doses a hankering after rest and sleep predominates."

III. *Hyoscine.* — This is the second alkaloid found in Hyoscyamus. It is also found in Datura. It is a syrupy liquid alkaloid. It is largely found in the substance spoken of above as *amorphous hyoscyamine.* In man Hyoscine produces dryness of the mouth, flushing of the face, and deep sleep, associated with semi-delirious muttering and giddiness. Mydriasis in usually, though not always, pronounced. The respiration is slow and full, and is sometimes of the character known as "Cheyne- Ṣtokes." The skin, so far from being abnormally dry, is often bathed in perspiration. It is asserted that there is a rise in temperature. There is sometimes paralysis of the pharynx, and of the muscles of the larynx. The mydriatic effect is associated with paralysis of accommodation and the maximum effect is produced in a third of the time required by atropine. Its influence in producing sleep is very marked. Nausea, constipation, and other disturbances of the stomach and alimentary canal are rarely witnessed. These remarks are based by Murrell on the authority of H.C. Wood of Philadelphia.

Just a word about Duboisine referred to in my foregoing observations. It is an alkaloid contained in Duboisia myaporoides, a tall shrub growing plentifully in the forest lands of Eastern Australia. It contains an alkaloid known as Duboisine identical with Hyoscyamine. The general action of Duboisine, says Murrell, is the same as that of Atropine. The former is so powerful, says Ringer,* that a 1 in 120 solution specially applied to the eyes often excites great giddiness, weakness, and a *drunken gait.*

IV. *Daturine.* - Lauder Brunton says that this alkaloid is a mixture of atropine and hyoscyamine†. He does not mention Hyoscine as a component of the Daturine-an "impure alkaloid" as he calls it. Fifteen years ago, Hyoscine was not determined. Ringer does not mention it in his Therapeutical work just referred to. Murrell even does not mention Hyoscine as a component of Daturine. In speaking of Daturine he only says "it is not a simple body," but a mixture of Atropine and Hyoscyamine. "The physiological action of Datura is identical with that of Belladonna, whilst Daturine has the same

* A Hand Book of Therapeutics, p. 539, 8th Ed., London, 1880.

† A Text Book of Pharmacology, Therapeutics, & c. p. 909, 2nd Ed., London, 1885.

action as Atropine, says Murrell. The symptoms of poisoning by Stramonium differ in no respect from those of poisoning by Belladonna. "The same accelerated pulse, the same elevation of temperature, the same wild delirium, the same increased frequency of respiration, the same widely dilated pupils, the same red efflorescence on the skin, the same restlessness or convulsions occur in both cases, and when the dose has been sufficiently large, and alike in abolition of the functions of circulation, respiration and innervation-stupor, general paralysis, weak rapid thready pulse, threatened asphyxia, constituting the phenomena of the closing scene in poisoning from either narcotic (H.D. Wood).

Having so far, and so freely dwelt on the pharmacological effects of the alkaloids found in Datura, it is time I considered the clinical evidence to be gathered from works on Medical Jurisprudence, chief among which stands the work of Norman Chevers. It is not my intention to dwell on the nefarious deeds of the infamous *Dhaturias* — who were known, in days gone by, as professional poisoners of travellers and tired wayfarers, poisoned because they were unsuspecting. The deeds of such professional poisoners will be best seen in Chevers, work. I therefore pass on to show instances wherein the above- -mentioned physiological effects were confirmed by various clinical observers, in cases of poisoning by Datura. Dr. Morehead of J.J. Hospital, Bombay, so far back as 1860 says, that in cases of Datura poisoning, the symptoms are in many respects allied to those of *delirium tremens.* The delirium is more muttering, not so busy as that of *delirium tremens;* but there is the same rambling of the mind on subjects not present to the senses. There is the same power of controlling the thoughts for a few moments, the same desire to appear rational, and above all, the same picking at small objects, as if they were indistinctly seen, which is often observed in the advanced stages of *delirium tremens.* Where the quantity of the poisonous stuff taken is large, there is coma with agitated movements of the hands and lips, and picking movement of the fingers : in fact, the same class of deranged nervous actions which characterize the third stage of *delirium tremens.* There is, however, this great difference in thses latter phenomena when caused by Datura-they are very generally recovered from, not by a return from coma to a state of health, but the coma ceases, and then succeeds the delirium, and the other phenomena which attend those slighter cases which have never passed into coma.

Dr. H. Giraud's papers on cases seen by him at the J.J. Hospital Bombay, namely one contributed to the Bombay Medical and Physical Society, and another cited in Appendix C. of Dr. Norman Chevers, Ind. Med. Jurisp. (p. 838, 3rd Ed., 1870, Calcutta) are well worth studying. Dr. Lyon quotes in his Medical Jurisprudence, Bombay, Dr. Giraud's remarks.

It must not be supposed that it is always that the Datura plant is used for criminal purposes; even when so used, it is generally not with the intention of causing death, although Norman Chevers cites cases where such was the intention. There are instances of persons having died from eating the leaves by mistake, while more than one little child has either died or suffered very severely from swallowing its seeds. (Anne Pratt). When the first settlers arrived in Virginia, says Sowerby, some ate the leaves of Datura and experienced such strange and unpleasant effects therefrom, that the colonists called it the 'Devil's Apple' name by which it is still known in the American States. In most cases it has been eaten by children in mistake for some other wild plant. Dr. H. Cleghorn, of the Madras Medical Service, notes a case, (quoted by Norman Chevers),* in which fragments of 3 or 4 leaves were found in the stomach of a poisoned Indian child 2 years old. "The mother of the child was reaping in a *raggy* field, when it was discovered that her two children were eating the leaves of Datura. A leaf was found convulsively grasped in the hands of one of them. It was the leaves of *D. fastuosa* that the children had been eating. One of the many species of Thorn-Apple possessing the poisonous properties mentioned above, is said to have been used in ancient days by the Priests of Delphi "to produce those semi-delirious paroxysms which they palmed off on the multitude as the results or manifestations of divine inspiration. The seeds of another species of Datura were similarly employed by ancient Peruvians." (Sowerby.)

I have at present under my care, in the Ratnagiri Lunatic Asylum, an inmate named Dhondnak Kannak. He admits to have been an inveterate smoker of Datura leaves, and believes he is an "inspired

* Page 194, Ind. Med. Jurisprudence, Ed. 3rd, 1870, Calcutta.

being". He never smoked the leaves for committing suicide as some persons are known to have eaten them, or the seeds for self-destruction. He has for a long time smoked the leaves to find relief from the asthmatic fits to which even now he is a subject as an inmate of the Ratnagiri Lunatic Asylum. Of course he gets no Datura smoke in the Asylum.

Professor Robert Christison, of Edinburgh University, says that the cases of poisoning from Thorn-Apple which occurred in his country (Scotland) up to the time he published his work on Poisons (1845), were all accidental. There are several such instances of accidental poisoning in other countries. Thus for instance, in America, in 1765, when some of the British troops under Sir John Sinclair were stationed in the vicinity of Elizabeth Town (New Jersey), three of the soldiers collected some quantity of the Datura plant which they mistook for the safe simple table-green food named *Chenopodium album*, dressed it, and ate it."* One of these soldiers became furious and ran about like a madman; the second was seized with genuine tetanus and died; what happened to the third is not mentioned. Beck on the authority of Orfila, cites a case in which a man after having been poisoned with the Thorn-Apple survived, and was cured of an intense long-standing headache. Even the bruising of the leaves of Datura in a mortar is known to have caused dilatation of the pupil. The application of bruised leaves of Datura on raw abraded human skin is known to have produced dangerous symptoms of poisoning. The empyreumatic oil of Stramonium is said to be poisonous to animals. (Beck.)

Taylor says that the seeds of fruit scarcely ripe are not very bitter, Children, therefore, eat them taking the fruit for some other fruit, not knowing its poisonous nature†. The seeds retain their poisonous properties notwithstanding exposure to heat. Mr. Lobo met with the case of a child aged five who ate more than a drachm of the seeds slightly roasted. In about an hour poisonous symptoms appeared. (Taylor on Poisons, p. 784, Ed. 1848). Dr. Chevers mentions a case of non-criminal poisoning from the leaves of Datura which is very remarkable. In March 1866,

* This case is recorded by Dr. Barton and mentioned in Beck's Med. Jurisp., p. 910, 1836, London.

† I was once very nearly eating the fruit myself when barley five years old with my little brother, aged three years, I took the Thorn-apple for custard apple. *K.R.K.*

somewhere near Midnapore, a whole family consisting of father, mother and children ate the cooked young shoots of Datura as '*greens*' by mistake, supposing them to be harmless. The parents died, but the children recovered.* The stomachs of the parents were sent to Calcutta for chemical examination. The analysis was scientifically tested on a puppy. The essence obtained from the stomach-contents of the deceased parents was injected into the stomach of the puppy. Vomiting set in immediately afterwards; very soon after that the puppy performed a few antics and fell over on his side into a deep sleep from which tickling failed to rouse him. He, however, recovered after six hours, but the pupils were exceedingly dilated and they continued more or less so the whole of the next day. Lucky puppy! He did not die!!

Chevers describes another case which is well-worth reproducing. In this case it is a kitten that is experimented on. In the Midnapore District, in September 1866, a sub-inspector of police died while he was in custody. It was reported he had taken opium. His stomach was sent to Calcutta for chemical examination. No opium-trace was found, but particles were found in the stomach which had the appearance of Datura seeds. A decoction of these particles and the whole stomach was made. and half of it was given to a kitten at noon on 27th September. "The little cat soon began to breathe with difficulty, and to froth at the mouth; in ten minutes her pupils were dilated, and they continued to remain so, only to a still greater extent, the rest of the day, never for a moment being contracted, or even less dilated even when exposed to a strong sun-light. After 20 minutes the kitten was placed in the middle of the room, and encouraged to walk but she staggered & fell attempting do so. In half-an-hour from the time of administration, she was quite unconscious; up to this period she felt pain when pinched with forceps, but now a severe

* Some years ago I had case of poisoning of a whole family under treatment at the Thânâ Civil Hospital. It was accidental, and not criminal. A father, mother with a baby in arm, and two other children were brought to hospital, reeling and staggering, trying to catch imaginary objects in air. The infant nursed by the mother did motions of hands as if to brush off something from its face. This shows that before the mother's milk was arrested the infant was poisoned. They all recovered. I have a photo-group of them, but it is not handy just now. (*K.R.K.*)

pinch only caused a slight movement of the limb, without any expression of pain. The respiration was laboured; she continued to froth at the mouth, and the pupils remained very widely dilated. Consciousness began to retun at two o'clock; she then got up, sat staring widly and commenced to perform a series of grotesque actions, ultering a slow moan from time to time. When pinched, she felt pain, but not yet very acutely. She appeared very irritable, almost wild, but was neither vicious nor bad-tempered. At 3 p.m., the pupils were dilated extremely, the iris being a mere thread. By 4 p.m., she had recovered so far as to come when called, and to feel acute pain when pinched, the pupils continuing as large as ever." (P. 190. Chev. *op. cit.*). Fortunate kitten! She did not die!! From this experiment, and from a similar one, confined only to the eye of another kitten, it was evident that the sub-inspector of police referred to above had died in the lock-up, not from self- administered opium as was suspected, but from Datura poisoning. Herein is the triumph of Western learning and scientific chemistry and chemical analysis over Eastern ways of crime and misadventure!

The motives for administering Datura seeds or leaves—seeds especially — are various. They do not necessarily mean a desire to kill the victim to whom they are administered in various forms,—in drink, in common bread, or sweetmeats. They are given for instance, says Chevers, to a shopkeeper, to rob him, when intoxicated, of his articles of merchandize; to a Fakir, to make him yield up the contributions of the pious; the drug again is given frequently through jealousy to secure revenge; it is likewise given out of pure fun. Dr. Ralph Moore, says Chevers, was once sent for suddenly to the jail Dr. Moore was in charge of (somewhere in Bengal), where he found the entire jail-guard scattered about on the floor, under the influence of Datura. As there was no attempt on the part of the prisoners to escape, it would appear that this wholesale intoxication the jail-guard was probably *a mere practical joke*, without any slightest intention of committing a crime. In one case mentioned by Chevers, the intention of administering Datura poison was not crime, but the mere desire on the part of a widow to stupefy three persons from whose company she wanted to escape! One of the three victims was a man who admired her for her good looks, but he did not knwo that she had sense enough to escape from one she had no regard for. I can add many more instances like those mentioned above regarding the non-criminal use of Datura, but I must consider the space I can reasonably claim in this journal.

Dr. Norman Chevers dwells a great deal on the criminal use of the Datura plant by the natives of India, namely, for stupefying the victims and then robbing them of their money, by the administration of either Datura leaves or pounded Datura seeds in bread, sweetmeats or drink. But I must add here that the practice of poisoning for criminal purposes is well known to have been followed even in European countries, as testified to by European writers. It only means that human criminal nature is alike, all over the world, where gain by sordid means is concerned. Dr. Christison has it, in his work on Poisons cited above, that the Thorn-apple in his day was being used in Germany to cause loss of consciousness and lethargy, preparatory to the commission of various crimes. It is known to have been used in France also for making men *insensible* with wine, in which *Thorn-apple seeds had been steeped.* Note that thereafter the insensible and help-less men were robbed of their personal effects. Vicat in his treatise on the Poisonous Plants of Switzerland mentions a case of poisoning by the Thorn-apple plant.

Dr. Chevers cites cases in which Datura is used for *suicidal purposes.* (See p. 203. *op. cit.*). Dr. Shortt of Madras has recorded that a middle-aged Brahmin returned as usual in the morning from his field, and fell, admitting, on being questioned by the women of the house, that he had eaten Datura leaves. Dr. Shortt says, that the suicide of this Brahmin was imitated by two girls who lived not far off, and who poisoned themselves with Datura while the sensation caused by his death prevailed. It is well-known fact that suicide is often a contagious infirmity of the human mind. Dr. Norman Chevers work is replete with many more instances of accidental or non-criminal, as well as criminal intentional poisoning with Datura. But I must say *Buss* after having so largely quoted him in the foregoing remarks.

Reprinted from
Journal Bombay Natural History Society
Vol. 15 : 56-69. 1896

PART XX.

JATROPHA CURCAS, *Linn.*

Reprinted from
Journal Bombay Natural History Society
Vol. 15 : 56-69. 1896

PART XX.

JATROPHA CURCAS, *Linn.*
NATURAL ORDER—EUPHORBIACEÆ.
MARATHI— जेपाल, (*Jepal*), मोगली एरंड (*Mogli Erand*).

A large, low-branching, irregular, scraggy shrub.

BRANCHES — irregular; younger ones with close-packed leaves and green bark.

BARK — of older branches from one to two lines thick, light-green, covered with a thin pellucid or translucent paper-like epidermis which cracks and falls off in irregularly-shaped pieces. The bark on section with a penknife pours out copiously a thin translucent whitish juice which on exposure to air turns brown or even blood-red, and dries up in opaque tears or irregular masses. The bark of the oldest branches is light-ash- coloured, and has here and there innumerable yellowish spots of the size of a millet-seed.

WOOD — white, very soft and spongy; pith well marked and dense in young and topmost branches. The wood, says Dymock, is loaded with starch.

LEAVES — scattered on the older branches; closely-packed at the top of youngest branches; alternate; the youngest leaves of a reddish tinge, with a tomentose under-surface, the tomentum disappearing as the leaf grows old and assumes a green colour. Upper surface smooth. Shape broad-cordate, or orbicular-cordate; 5-angled. Hence called "*Angular leaved Physic-nut*". Hooker says the leaves are 3-5 lobed, but as generally observed they are distinctly 5-lobed. Lobes 4-6 inches in diameter.

NERVES — otherwise called VEINS, well marked and prominent on the under-surface.

PETIOLES — as long as the blade; round; smooth; inserted on a kidney-shaped projection of the bark and surrounded at the insertion by a thin expansion of the epidermis. STIPULES — absent.

INFLORESCENCE —

Flowers moncious, yellow, or yellowish green; pubescent; in terminal corymbose cymes bearing many small flowers; the central flower in the cyme or in its forks being always female. The flowers are scentless absolutely. The peduncle is often seen arising from exterior axils. BRACTS, entire small, one below each sub-division of the cyme, and generally one pressing on the calyx.

I. MALE FLOWERS —

Found at the extremities of the ramifications of the main flower stalk, on short articulated pedicels; deciduous.

CALYX — greenish; sepals five, lanceolate, imbricate.

PRÆFLORATION, convoluted.

COROLLA — Yellow; ¼ inch in diameter; petals five, ovate- oblong; villous within. Hooker says that the corolla is salver- shaped; Roxburgh says it is companulate. It is the female corolla that is campanulate, as will be seen from my Plate 'V' accompanying.

PRÆFLORATION — contorted.

STAMENS — many, says Hooker; 8-10, says, A. de Jussieu; 6 says Roxburgh.

FILAMENTS — connate, says Hooker. A. de Jussieu describes them thus — "Inferiorly connate, of which 3-5 are interior and larger than the outer ones." (Euphorbiacearum Genera, p. 37, Paris, 1824). Roxburgh, on the other hand, describes the arrangement of the filaments thus — "Filaments *six*, the central one very thick and columnar; the five external ones filiform, towards the base adhering to the central one; all erect and a little longer than the calyx." The central mass of filamentous column represents the abortive Gyncium

ANTHERS – yellow; often tinged brown. Says Roxburgh – "Ten, sagit tate, equal; five supported by the long central filament, and one by each of the others." Hooker describes the anthers as erect, 2-celled. The anther-lobes dehisce longitudinally from above downward. Anthers, says Baillon, are extrose, dehiscing marginally. POLLEN – large, spherical. DISK – always present but varies in form. It is either a yellow entire ring, or is made up of five yellow glandules or squamules "The glands of the hypogynous disk are free or united" (Baillon).

II.- FEMALE FLOWERS—

Seated in the divisions of the male pedicels, "with their own pedicels *not articulated*" (Roxb.).

CALYX — as in the male flowers, but persistent.

COROLLA — as in the male flowers, but more substantial, and distinctly companulate.

DISK — not so well-marked as that of the male flowers.

STYLES — 3; bifid or bi-lobed (A. de Jussieu); short. "Connate below or united on a slender column; glabrous" (Hooker).

STIGMAS — 3; bifid, pubescent.

OVARY. - 3-carpelled, each carpel containing one ovule; superior "glabrous" says Hooker. Note that the ovary and styles are said to be pubescent in *Jatropha curcas* of Wallich (Syn., - J. Wightianus), — Hooker. It may be noted here that at times only one ovule, or two, mature into seed. Very seldom have I seen the three ovules yielding three seeds. But that the three ovules do so yield three seeds will be seen from my illustrative plate accompanying this paper. The ovule is pendulous from the inner angle of the central column.

FRUIT — A tricoccous capsule; oval, or globosely oblong; $1-\frac{1}{2}$ inch long; ¾ inch broad, or even 1 inch; green when first formed; yellow when mature; brownish or black when dry. Hooker says there are 2-4 cocci, but I have never seen the fourth coccus.

EXOCARP - thick and succulent; fully of milky or pellucid white juice; 2 lines thick. Black or brownish, & easily removable when dry.

ENDOCARP — "crustaceous or bony" says Hooker.

SEEDS — $\frac{1}{2}-\frac{3}{4}$ inch long; $\frac{3}{8}$ inch broad; oblong; laterally attached about the middle of the cell, with a distinct white aril at the hilum; convex on the outer or dorsal side; angular on the inner or ventral side.* There is a well-marked ridge on the ventral surface (Dymock).

TESTA — smooth, black-crustaceous.

"The *lorica* and *tegmen*", says O'Shaughnessey, "are separable. The internal *tegmen* is hard, brittle, with resinous and brownish fracture; the almond is white, foliaceous, covered with a thin white membrance." This membrane is like tissue-paper and easily separable. The almond is inodorous, of agreeable taste, but slightly acrid after

* O-Shaughnessy's Bengal Dispensatory, p. 558, Calcutta, 1841.

chewing. It renders the saliva emulsive, and irritates the fauces (O'Shaughnessey).

ALBUMEN — "Fleshy" says Hooker; "Copious and oily" says Dymock (Pharmacographia Indica, Vol. III., p. 276, Bombay 1893). It is best to speak of the "Albumen" as *Endosperm.*

COTYLEDONS — broad; flat; foliaceous; adhering closely to the large mass of endosperm. The endosperm cells contain a net-work of protoplasm saturated with oil. This oil does not exist in the form of globules or drops, though it can be extracted by pressing the endosperm-tissue as in the case of the seeds of the castor- oil plant.

RADICLE — superior, short, thick.

EMBRYO — straight, central.

GENERAL REMARKS

Jatropha curcas is an exotic, naturalized in India. It appears to be a free and rapid grower in this country, not wanting much care. It is a hardy plant, which has taken quite kindly to the soil of Western India whether it be in the Konkan or in the Dekkan. In both these divisions of Western India, I have seen it grow profusely as a hedge-plant, where no human hand has watered it. It evidently takes its nourishment from the air, and from the soil in which it grows, depending mainly on the rain-water and dew, whenever it can get it. In the Konkan it gets its water- supply from the monsoon rains from June to October. Hooker says that the plant is ever-green. It is not so in the Konkan. I have seen that in the Thana and Ratnagiri districts it is leafless, though in inflorescence during April and May. Nay, in 1898 in Satara (Dekkan) I found the plant leafless in January and February. The plant is a native of Brazil and of the West Indies. From Brazil it is said to have been introduced here by the Portuguese (Dymock). It may be so; it is probably introduced here from Northern Africa also, by way of Arabia. No mention appears to have been made of this plant by Henrique Van Rheede in his elaborate work entitled *Hortus Indicus Malabaricus* published at Amsterdam in 1678 in six folio volumes, fully illustrated with the aid of three local Pandits-Rango Bhat, Vinayak Pandit, and Appa Bhat. The Marâthi or *quasi- Marâthi* letter-press of the preface these three Pandits then wrots is very characteristic of the Marâthi spoken and written in Malabar in the seventeenth century.

Jatropha curcas, however, appears to have been mentioned in Bhâv Prakash, a Sanskrit work describing several Indian plants. "Its author, Bhâv Misra," says Dattaram Chobhe, "flourished in Madras about 350 years ago." Surgeon-General Balfour of Madras says that Bhâv Misra flourished so far back as 1550 A.D. It is possible then that the Portuguese must have introduced the plant into India. Sir George Birdwood says that "Jatropha curcas was first mentioned by *Monardes*" (Veg. Products, Bombay Presidency, page 77 and page 308, 2nd ed. Bombay, 1865). *Monardes* is evidently a misprint for *Monardus*, for no such name as *Monardes* is mentioned by Professor Sachs in his History of Botany (1530-1860), published at WÜrzburg in 1875, and translated into English by Garnsey and Professor Balfour in 1890 (Oxford). I find, however, from John Gerarde's *Herball* that Thomas Johnson, an Apothecary of London who re-edited Gerarde's Herball in 1663 A.D., mentions an American writer named Nicolas Monardus. Johnson, in addressing his readers in a prefatory note, speaks of Nicolas Monardus as being a writer on the simple medicines of the West Indies. Evidently Monardus flourished about the middle or end of the sixteenth century. From Johnson's remarks in Gerarde, I find that the works of Monardus, originally written in Spanish, were translated into Latin by Carolus Clusius between 1583 and 1601. "Carolus Clusius", says Johnson, "was a learned, diligent, and laborious *Herbarist*." Carolus Clusius, a Frenchman by birth, was otherwise named Charles de I'Ecluse. He was born in Arras in 1526. His family suffered from religious persecution in France, and he spent the greater part of his life in Germany and the Netherlands. In 1573 he was invited to the Imperial Court of Vienna by Maximilian II. Clusius accepted the invitation and removed to Vienna. Subsequently in 1593 he became Professor of Botany in Leyden and died there in 1609 (Sachs).

O'Shaughnessey notes that Jatropha curcas is a native of New Andalusia & Havana (in the Island of Cuba). A. de Jussieu says that it is also a native of North Africa. Now, it is a well-knwon fact that India, especially its Western Coast, and North Africa, through Arabia, have for several centuries past been in close mercantile intercourse with each other. It is, therefore, just possible that Jatropha curcas was introduced into India by the mercantile Arab & Africa visitants of the shores of Western India. But I have no authentic information on this point. It is a mere surmise of mine. The Africa coast has given to the

Ratnagiri Coast a class of fishermen called the *Dâldis* who are distinctly of Afric blood and origin. Possibly their progenitors brought the plant with them. Why not ?

In English parlance the plant is known as the *Physic-nut* or the *Purging-nut*. It must be noted that it is quite different from the plant known as *Poison-nut* which is Strychnes Nux-Vomica. The seeds of Jatropha curcas are known as *Pignons d'Inde* (Dymock). The oil obtained from the seed is spoken of as *Oleum infernale* by Professor Dr. L. Lewin of Berlin (page 374, Lehrbuch der Toxikologie, 1897). The oil is similarly spoken of by Dymock thus — "It was formerly employed as a purgative by European physicians under the names of *Oleum Ricini majoris* and *Oleum infernale*." Roxburgh says that the plant is one of the most common on the Coromandel Coast, and in flower and fruit all throughout the year. This fact is worth noting, as showing the effects of local influences on the growth of plants. Talbot of Canara fixes a definite time for flowering and fruiting. That of flowering from April to May; that of fruiting during the rainy season. I must note here one important remark which Roxburgh makes regarding the wood of the plant. In observing that the wood of *Jatropha curcas* is too soft and spongy to be of any use he cogently adds that the wood "will not even burn freely."

With regard to the oil expressed from the seeds, O'Shaughnessy and Roxburgh say that it is used to burn in lamps by the poorer classes of the natives of Bengal. I am not aware that it is used on this side of India for any such purpose. But in America it appears to be so used freely.

In noticing this plant as an exotic in his Flora of Ceylon (Vol. IV., p. 46, London 1898), Dr. Trimen says that it is very commonly planted as a fence round native gardens, and that it is a violent purgative.

Dymock says that *Jatropha curcas* is said to have been introduced from Brazil by the Portuguese. Sir George Birdwood gives South America as the *habitat* of the plant. The plant seems to be well-known in China, as the Chinese turn out a fine black varnish by boiling the seed-oil with Oxide of Iron. Dr. Dymock observes that the juice of the plant, when dried in the sun, forms a bright reddish-brown, brittle substance like shell-lac, which may be put some useful technical purpose (Pharmacogr. Ind., Vol. III., p. 275). A similar remark is made by Dalzell and Gibson - "The fresh juice of the stem when dried forms an elegant lac-like substance, which may yet be applied in the Arts" (p. 77, Supplement, Bombay Flora, Ed. Soc. Press, 1861).

Though the oil of the seeds of *Jatropha curcas* is largely used in lamps in the different parts of India mentioned above, it is known in European commerce as the "seed-oil" (G. Birdwood). No such oil that I am aware of is exported from India. So says Sir George Birdwood also.

In a work entitled "A General System of Botany" by Emm Le Maout and J. Decaisne, translated by Mrs. Hooker in 1873, London, Longmans, Green & Co., at page 698, I find the following remark which is well worthy of the consideration of Indian soap makers. The remark runs thus — "The seeds of *Jatropha curcas* [Physic- nut], a shrub growing throughout the hot countries of America, yield a plentiful supply of an oil of which soap is made."

Garcias de Orta, in his Colloquies on Indian Plants, mentions a Plant named *Jatropha* (see p. 79, 2nd Edition, Lisbon, 1872). Whether it is the same as *Jatropha curcas* or not I am not able to determine.

The following are the Synonyms of Jatropha curcas, *Linn*, as given in Hooker's Index Kewensis (pages 1251-1252, Part II, 1893, Clarendon Press, Oxford) —

1. Jatropha acerifolia, Salisb. Prod. 389.
2. J. condor, Wall. Cat. N. 7799.
3. J. curcas, Wall. Cat. N. 7799 D = Wightiana.

Note that there is a plant called Jatropha peltata, figured by Wight in his plate 1169 in the Icones Plantarum which is the samc as jatropha Wightiana referred to above under synonym No. 3. Note again that what Wight depicts as Jatropha villosa in plate 1159 of his Icones is said by the Compiler of the Index Kewensis to be nothing but Jatropha Wightiana. The older writers speak of this plant as Curcas purgans. Back says it is the *Tuva tree* of the Philippine Islands. It is also known as *Indian nut*.

In a list of five hundred Indian Plants published in Canarese at Mangalore by the Basel Mission Book and Tract Depository under the authorship of C. Stolz, the following synonyms are given —

(1). Castiglionia lobata, R.B.

(2). Ricinus americanus, Mill.

(Page 11, 2nd Ed., 1891).

It may be noted here that like the leaves of the Castor-oil plant (Ricinus communis, *Linn.*) the leaves of Jatropha curcas have galactagogue properties. A decoction of the leaves is used in the Cape de Verd Islands to excite secretion of milk in women (A.A.B. in Maunder's Treasury of Botany, Part I, page 363, Edition 1870). Dr. Bennett of Sydney (Australia) is credited with havign made the following observation in his work entitled *The Gatherings of a Naturalist* —

"The milky acrid glutinous juice, when dropped on white linen, produces an indelible stain, at first of a light-blue colour, but after being washed changes to a permanent brown : it might, therefore, form a very excellent marking ink." I hve not been able to obtain such a stain. Will any of my readers help me in settling this point ? The oil of Jatropha curcas seeds is said to be of a light colour, and used as a substitute for Linseed oil, as well as for dressing cloth (Maunder's Treasury of Botany). It is also said to form a basis for the red dye of the cotton fabric knwon as *Turkey red.*

The following remarks are made with the object of engaging the attention of those earnest students of plant-life who are interested in the investigation of the question as to how the environment of a plant in the Eastern and Western hemispheres respectively affect the existence and growth of any particular genus or species of this or that plant. It is well-known that several members of the Euphorbiaceae, like several members of the Cactaceæ, are succulent plants. Strictly speaking, Jatropha curcas is not a succulent plant, but at all times of the year in Tropical East India, especially on the coast of Western India, in the Konkan and in the Dekkan, it is rich in a milky juice. Such plants, rich in milky juice all throughout the year, may during their existence in the hot weather be aptly compared to camels — as the "ships of the desert." Indeed, they provide for themselves a large quantity of water and are then able to dispense with further supplies for a long time without injury. "The cells of the aqueous tissue are comparatively large, and their walls thin; the active protoplasm within forms a delicate layer round the walls — that is to say, a sac whose cavity is filled with watery, often mucilaginous fluid." These remarks of Kerner Von Marilaun's (Oliver's Translation, History of Plants, p. 328, Vol. I., 1894, London) may well apply to Jatropha curcas

It may be noted here that the flowers of Jatropha curcas besides being odourless dont appear to invaded by insects. Cattle wont touch its leaves. Hence as a hedge-plant it is very serviceable as already noted in my foregoing remarks. I may here emphasize the fact that Jatropha curcas is one of the Euphorbiaceous plants in which each plant without exception develops both true staminate and pistillate flowers on one and the same flower stalk. It must be further observed that the members of the Euphorbiales, not to speak of the genera only but also of the specimens of the species and varieties locally obtained, differ at times very widely in individual development of parts of the plant under local influences of soil and climate. But what I have noted above is the result of my personal observation of this plant in its living condition in Bombay, Thana, Satara, and Ratnagiri.

I have noted above that Dr. Dymock speaks of the seed of Jatropha curcas as *Pignon d'Inde.* Just a word of warning to my readers. I find from Baillon's Natural History of Plants, translated from his elaborate French work by Marcus Hartog, that there are two kinds of *Pignon d'Inde, viz.* — the large and the small. The large one is the seed of Jatropha curcas, otherwise called by Baillon as the Jatropha of Barbadoes (Castiglionia lobata, *R. et Pev.*). The small *Pignon* of India is the seed of Croton Tiglium, the oil of which is much more deadly (p. 162, Vol. V., 1878, London. Baillon's Translation into English by M. Hartog).

From Paxton's Botanical Dictionary it appears that Jatropha curcas was introduced into England from South America in 1731. It is noted as an ever-green plant (p. 310, Edition by S. Hereman, 1868, London).

Baillon remarks that the wood of Jatropha curcas is "soft, easily decaying; palisades are however made in the Indies" (West — K.R.K.). Baillon further adds that Jatropha curcas is used for hedge rows in the Antilles (Marcus Hartog's English Translation of Baillon, Vol. V., p. 175).

POISONOUS PROPERTIES.

The seeds of *Jatropha curcas* are poisonous. They produce, when taken internally, vomiting and drastic purging. The oil obtained from the seeds has similar effects. "The seeds are employed by the native doctors of the Philippine Islands and considered excellent and mild purgatives in doses of from one to four seeds. The effects which result from an overdose are vomiting purging, a burning sensation in the stomach, with a determination of blood to the head. * * * Dr. Bennett, the well-known naturalist of Sydney, has himself administered these seeds to Europeans, but has found their effects very irregular, and occasioning, in all cases, a burning sensation in the bowels followed with nausea and vomiting" [A.A.B. in Maunder's Treasury of Botany].

Beek says that the effects of seeds are at times so great that in addition to vomiting and purging there is insensibility, and great weakness followed by death. Violent inflammation of the mucous membrane of the stomach and intestines has also been noted (p. 832, Medical Jurisprudence, 5th Edition, London, 1836).

The following is an extract from Dr. Hilton-Fagge's Report on Toxicology and Materia Medica (p. 347, New Sydenham Society's Year-Book, 1865, London) — "In the *Medical Times and Gazatte* — (London, 1864, I. 703) — are recorded several cases of poisoning" by the seeds of *Jatropha curcas*, "which occurred at Birmingham, where three or four sacks of the nuts were sold by auction. These nuts were left on the floor, and some boys, getting access to them, put some into their pockets, and finding their taste pleasant not only ate some of them themselves, but gave others to their friends. In consequence thirty-three persons were taken so ill as to be obliged to be carried to the General Hospital. Some of them had only eaten three or four of the nuts, but others ate as many as fifty. Symptoms came on in some cases in ten minutes; in others not until two to two hours and a half had elapsed. The chief effects were pain and burning in the throat, pain and distension of the abdomen, giddiness, vomiting and drowsiness; and after an interval, purging-the evacuations being copious, mucous, and not unlike the rice-watery stools of cholera. In some cases the depression was very great; in many there was dysuria; fever was always present as an after-effect. Dilatation of the pupils was believed to exist in those who took a large quantity of the poison. It is doubtful whether the drowsiness-from which the patients were easily roused-was a narcotic effect of the nuts or whether it was caused by the great prostration." [*N.B.* — I am inclined to think the latter caused drowsiness — *K.R.K.*]. Strangely enough, all the cases recovered ! The treatment consisted of the administration of emetics and purgatives. Will not the Homopaths says "*Similia similibus curantur?*"

Dr. Norman Chevers* quotes a case reported by Morrett of Salem (Madras Presidency), and fully detailed in the Madras Quarterly Journal of Medical Science, p. 37, July 1861. The case was that "of a healthy young English gentleman, who ate fifteen or twenty nuts of *Jatropha curcas*. The taste did not arouse suspicion. In an hour and-a-half burning sensations in the throat and stomach were felt, attended with great restlessness. There was profuse purging, with great pain and vomiting, by which portions of the nut were rejected. In another hour and-a-half there were severe cramps of the lower limbs. The purging now ceased. The cramps were so violent that frequently the patient sprang in agony out of bed, and stamped his feet with all his force upon the ground as if to flatten them, the cramps distorting them. The cramps affected, with less severity, the muscles of the arms, abdomen and back. Occasionally there were convulsive twitches of the back, almost like those in tetanus. He gradually became perfectly deaf. There was also impairment of sight. The pulse was very small, thready and weak. Skin cold, damp, and clammy. Features sunken and contracted. *Treatment* — Hot brandy and water, warm bath, anodyne frictions, and hot bottles. In about seven hours reaction set in and he slowly imporved. Throughout the following day he remained totally deaf, and had slight cramps. Next day he was awakened by the church bell, and was quite convalescent. Although during his illness he had appeared quite conscious, he had no recollection whatever of what happened later than the profuse purging. Several other young gentlemen were in like manner affected in a less alarming degree, having taken fewer of the nuts, but in none of them was there deafness or loss of memory." In giving this long extract from Chevers my only apology is that his work is out of print now. My quotation, therefore, is likely to be of use to those who do not happen to have his work in their library.

So far as I know, the seeds of *Jatropha curcas* have never been used for criminal poisoning. Brigade-Surgeon Lyon, C.I.E., in his Medical Jurisprudence (p. 205, Ed., 1889, Calcutta) says that several cases of *accidental poisoning* are recorded. My foregoing observations will amply bear him out in his assertion.

* Medical jurisprudence for India, p. 275, Calcutta, 1879. Thacker, spink & Co.

To add further evidence with regard to the poisonous nature of *Jatropha curcas* I quote Sir Robert Christison of Edinburgh. He observes that the seeds of the physic-nut of the West Indies, "when applied in the form of powder to a wound, produce violent spreading inflammation of the subcutaneous cellular tissue; and when introduced into the stomach they inflame that organ and the intestines. Four seeds will act on man as a powerful cathartic. I have known violent vomiting and purging occasioned by a few grains of the cake left after the expression of the fixed oil from the bruised seeds; and in some experiments, I performed a few years ago, I found that from twelve to fifteen drops of the oil produced exactly the same effects as an ounce of castor oil, though not with such certainty" as that of castor oil, I presume. *[K.R.K.*]*.

I offer to my readers the following quotation from A.S. Taylor's Medical Jurisprudence (2nd Ed., Vol. I., p. 328) —

"In August 1858, 139 children in Dublin suffered severely from eating some of these nuts, namely, of *Jatropha curcas (Medical Times and Gazette*, August 1858). They all recovered."**

"M. Chevalier refers to a case in which thirty-three persons were poisoned by eating these seeds. The symptoms which they suffered from were nausea, vomiting, and general depression. Twenty were so ill that they were placed in the beds of an hospital; the remaining thirteen soon recovered."

In 1871, when a student in Grant Medical College, Bombay, I ate two seed almonds of *Jatropha curcas* which was then growing near the Vehar water pipe in the College garden in front of the Clock-tower. I suffered from the effects of the almonds for fully six hours, though the nuts were pleasant to eat. My friend and fellow-student Khan Bahadur Dr. K.B. Cooper, now Civil Surgeon of Shikarpur, also ate some with me. Perhaps he will remember the incident. I remember it very well, as the emesis it produced was very trying. My medical attendant, Dr. Sakharam Arjun, of sacred and loving memory, then living in the same house with me, gave me a word of warning not to play pranks with my life in the course of my botanical studies by tasting unknown plants, merely for the sake of experiment. Such foolhardiness, I know, has often destroyed valuable lives. There is a saying in Marathi जेपालची मात्रा ॥ वैकुंठाची

*Christian's Treatise on Poisons, p. 591, 4th Edition, Edinburgh, 1846.

which means that if *jepal* is recklessly used, death is the result (Mr. G.S. Bhide, Jailor, Ratnagiri Prison).

As regards the seat of the poisonous element, O'Shaughnessy says that it is the *embryo* which is the source of the acridity of the seed and its poisonous nature. I doubt this very much. I feel almost sure, that the acridity and the poisonous principles lie in the once so-called albumen, and now called the endosperm. I do not at all believe, as suggested by O'Shaughnessy, that the removal of the little, insignificant embryo rids the nut of its poisonous property and makes it "bland and alimentary." Although this statement is made by O'Shaughnessy on the high authority of M.M. Feé and Humboldt, I must set my face humbly against it.

I find myself "all at sea" with regard to the chemical nature of the poisonous element of *Jatropha curcas*.

The active principle of the oil of *Jatropha curcas*, says Râi Bahadur Kanny Lall Dey, F.C.S. of Calcutta, has been named *Jatrophic acid*. A recent research (A. Siegel, 1894) attributes the activity of the seeds to a poisonous *toxalbumen* analogus to *Ricin* and named *Curcin* (p. 169, Indian Drugs, 2nd Edition, 1896, Calcutta). *Ricin*, says J. Charles E. Sohn, F.C.S., a member of the Society of Public Analysis of London, is a poisonous principle of castor seed of the *albuminous* kind (p. 191, Index to a Dictionary of the Active Principles of Plants, 1894, London). Dymock observes that Dr. H. Stillmark has discovered in the seeds of *Ricinus communis* an albuminoid body which has been named *Ricin*. This, however, says Dymock, does not appear to be the purgative principle (p. 277, Ph. Ind., Vol. III). It may be noted that *Ricin* appears to have a peculiar effect on blood, causing a rapid conglomeration of the red corpuscles, together with the formation of a substance like fibrin.

The oil of *Jatropha curcas* is said to contain an active principle similar to that of the castor seeds. It is known as Ricinoleic acid. Its formula is $C_{18}H_{34}O_3$. It occurs as a glyceride in castor oil together with *tripalmatin and tristearin*. It also occurs in *Jatropha curcas*. It is a thick only liquid which solidifies below O°C., and mixes in every proportion with alcohol and ether. Its alcoholic solution has an acid reaction, an unpleasant persistent acrid taste, and does not oxidize in the air (Roscoe and Schorlemmer's Treatise on Chemistry, Vol. III; Organic Chemistry, Part II, p. 484, Edition 1890, Macmillan).

I must not omit to mention here what Professor Schmiedeberg of the University of Strassburg says regarding the active principle of Castor oil. He says — "It is soluble in the intestinal fluids only; here alone it meets with the conditions necessary to its efficiency after it gets into the bowels. Like Croton oil, Gamboge and Jalap resin, Castor oil is insoluble in watery fluids, and consequently passes the stomach unchanged. In the intestines it is dissolved by the alkalies of the bile and pancreatic juice."* These remarks may well apply to the seed-oil of *Jatropha curcas*.

*Elements of Pharmacology, translated by Dixon of the University of Sydney, pp. 108 and 109 Edinburgh, 1887.

INDEX OF PLANT NAMES